"*Green Card STEM Voices: Stories from Minnesota Immigrants Working in Science, Technology, Engineering, and Math* is a powerful book that will inspire young people to pursue careers in STEM. These stories of Minnesota immigrants' journeys to their careers serve as a roadmap for all of our students—whether they are immigrants or not—on overcoming adversity and pursuing their dreams."

—Danielle Grant, President & CEO, AchieveMpls

"As a licensed middle school science teacher, former principal and currently working as a STEM consultant in and around the Twin Cities metro area, I was excited to hear about *Green Card STEM Voices*. It's diverse ideas and voices that will help this country and the world shape the future using the positive power of STEM."

—Dr. Tyronne Carter, Director of Marketing &
Curriculum Design for America's Fun Science

"This book is a powerful resource in restoring absent narratives of our immigrant voices in STEM professions. We must continue to honor the work of those who have dared to make the journey, physically and culturally, in the United States and celebrate the diversity of their stories and perspectives. The vision, grit, and passion tempts our current students to dare beyond their situations and to imagine a world filled with diversity. This book is groundbreaking in highlighting a world of STEM co-constructed through the immigrant experience."

—Dr. Aimee Fearing, Chief Academic Officer, Minneapolis Public Schools

"Seeing yourself in a potential career is how most young people create a vision of their own future. When most of the scientists we see in the media are not immigrants we are leaving out an entire group of future scientists. *Green Card STEM Voices* helps close that gap. Sharing the stories of how immigrants to the United States have forged careers in STEM fields will help young people see a different future for themselves. *Green Card STEM Voices* is a great tool for anyone working with youth and young adults."

—Shaina Abraham, Program Director, City of Saint Paul, Right Track

"America's greatest strength is our diversity, which is large and rapidly increasing because of immigration and children of immigrants. This dynamic diversity is particularly good news for STEM, because science is better when the scientists come from different perspectives and places. *Green Card STEM Voices* offers an important opportunity to hear the stories of recent immigrants who chose to work in STEM disciplines. By listening to these colleagues, we as a nation can learn and advance towards greater inclusivity."

—David J. Asai, Senior Director, Science Education,
Howard Hughes Medical Institute, Chevy Chase, Maryland

"Reading *Green Card STEM Voices* helps us understand some of our newest community members' lives. They bring perspectives that can improve the Science Museum of Minnesota experience that will enhance the experience for everyone. We need to look at ourselves—our programs, exhibits, hiring practices, and culture. Are we a welcoming space to work in and to visit?"

—Alison Rempel Brown, President & CEO, Science Museum of Minnesota

"*Green Card STEM Voices* has done it again—a unique new book and multimedia project highlighting the stories of our local Minnesotan STEM Superheroes. Green Card Voices shares eye-opening accounts of our refugee and immigrant communities in STEM. An intriguing read for current and future STEM professionals."

—Erin Twamley, STEM Educator and Author
(Everyday Superheroes: Women in STEM Careers)

"Beyond enriching our lives and our communities, one of the greatest benefits of harnessing and leveraging diversity lies in solving complex problems for the benefit of society. This book shows that coming to life in STEM-related fields. Time and time again, the great work of Green Card Voices educates all of us on the impact of cultural diversity and the role our immigrant populations play in making us a better society."

—John Guttery, Sr. Vice President, Enterprise Initiatives, Ecolab

"Green Card Voices has done it again, this time focused on STEM. In libraries, we know that stories are a powerful way to build connections, understanding, and empathy. We value the *Green Card Voices series* because it documents our immigrant communities' stories and shares them for the benefit of us all. This volume is another great addition, and we will be sure to add it to our library's collection."

—Catherine Penkert, Library Director, Saint Paul Public Library

"I am an immigrant who moved to the United States at a very young age, became naturalized, and am now a US Citizen. My childhood was in Beaver Dam, Wisconsin, where my siblings and I were among the only immigrants in the entire town. My father was a physician who cared for many people in the community and thus we were respected in our hometown. As I have grown older, I have learned that in order to change people's opinions of immigrants you must do what my father did: go one by one, helping people understand that the immigrant who lives next door might be the physician who saves their life, or preserves their quality of life, in their darkest hour. I am now a neurosurgeon and neurotrauma researcher, and my daily job enables me to help people, changing their opinion of immigrants one person at a time."

—Uzma Samadani, MD, PhD, Associate Professor,
Bioinformatics and Computational Biology, University of Minnesota

"These stories reminded me of two of my favorite words: curiosity and empathy. Everyone has a story that is worthy to be shared. I am encouraged to treasure my story and that of others around me. Embracing stories of others with curiosity and empathy makes our community flourish."

—**Ayumi Stockman, World Languages Content Specialist, Minneapolis Public Schools**

"The work of Green Card Voices has inspired me many times and this new book focusing on STEM Voices' is another extraordinary example of the power of their vision. My father was an electron microscopist for the US Department of Agriculture, and I trained in biophysics and biochemistry at Iowa State University. Respect for scientists was a core belief in our family and it was global-everyone on the planet had something to contribute to the creation of a better future for all through discovery and innovation. This book presents amazing stories that underline the universal nature of the human endeavors we call science, technology, engineering, and mathematics and these stories powerfully reminds us all of the importance of keeping our society open and welcoming in order to ensure a sustainable future."

—**Mark Ritchie, President, Global Minnesota**

"Green Card Voices has hit another home run with their latest book that captures the stories of immigrants working in the STEM field. They continue to keep the rich cultural traditions of storytelling alive. This book, without fanfare, shows not only the diversity of Minnesota's immigrant community but also the breadth of STEM related careers and is sure to be an inspiration to those who read it."

—**Shelley Andrew, Manager, Business Systems, Minnesota Twins Baseball Club**

"These stories are at once revealing and familiar, a testament to the incredible abundance that exists amongst our immigrant and refugee populations here in the Twin Cities—and what happens when that abundance is unlocked. They offer a window into immigrant lives, their struggles, and their future careers that were once near impossible to envision. But at the same time, these stories are relatable. They're grounded in the human experience—what unites us in this quest to build a meaningful life. Put this in the hands of every aspiring young person, whether they're involved in STEM or otherwise. It will help them see that, in fact, anything is possible—and to be who you are, anywhere you go."

—**Daniel Wordsworth, President, Alight (formerly American Refugee Committee)**

"The best way to understand other people's struggles and hardship is by connecting through their real life experiences. This book did an amazing job addressing those struggles and shed light on diversity in STEM careers. This book also provides awareness through personal stories."

—**Farah Faruqi, Ph.D student in STEM education, University of Minnesota**

Green Card STEM Voices

Stories from Minnesota Immigrants Working in Science, Technology, Engineering, and Math

Fadumo Yusuf, Thai Chang, Aasma Shaukat,
Getiria Onsongo, Raul Velasquez,
Dalma Martinović-Weigelt, Hussein Farah, Kim Uy,
Itoro Emmanuel, Elaine Black, Manish Shahdadpuri,
Karina Boos, Sampson Abiye Linus, Valerie Ponce,
Ingrit Tota, Fernán Jaramillo, Simeon Ngiratregd,
Apostolos P. Georgopoulos, Esther Ledesma,
and Zurya Anjum

Authors

Tea Rozman Clark and Julie Vang
Editors

ISBN 13: 978-1-949523-14-0
eISBN 13: 978-1-949523-15-7
LCCN: 2019920949

Printed in the United States of America
First Printing: 2020
20   19   18   17   16   5   4   3   2   1

Edited by Tea Rozman Clark and Julie Vang

Cover design by Nupoor Gordon
Interior design by Shiney Chi-Ia Her and Ellis Sherman

Photography, videography by Media Active: Youth Produced Media
Photography by Shiney Chi-Ia Her

Green Card Voices
2611 1st Avenue South
Minneapolis, MN 55408
www.greencardvoices.org

Consortium Book Sales & Distribution
34 Thirteenth Avenue NE, Suite 101
Minneapolis, MN 55413-1007
www.cbsd.com

We dedicate this book to immigrants and refugees,
especially Aleksandre Sambelashvili,
who have used their ingenuity and care to make lasting
impacts in the United States (and the world): healing and
saving lives, driving scientific and technological innovation, and
reimagining both everyday and extraordinary possibilities
for our people and the environment.

Table of Contents

Foreword

Although the story of my life evolves with each tick of the clock, I have told it many times. After years of sharing my story of coming to the US with audiences in schools, non-profit centers, and corporations, I was inspired to write and publish my memoir, *America Here I Come: A Somali Refugee's Quest for Hope.*

Like many of my fellow Minnesotans, I was born in Somalia. When the Civil War broke out in 1991, my family fled to the Dadaab Refugee Camp in Kenya. After three years living in refugee camps and navigating the grueling immigration process, my family resettled in the US—first in Denver and later in San Diego's City Heights neighborhood. My experiences as a refugee have profoundly shaped my career as I strive to rewrite the social and economic rules that exclude people from opportunities.

My story is both personal and one shared by many who have come to this country escaping horrible conditions at home or who have made decisions to continually persevere despite adversity. Through the process of writing my story, I have gained a deeper understanding of myself and the people around me. Sharing it has connected me with many amazing people who said they found my story to be a source of inspiration.

In all of my work, my mission is to provide leadership and to create opportunities for a better world, especially for refugees and those who are most vulnerable in our society. Telling my story has been a part of that work too. While *America Here I Come* was a memoir, I also wrote it to be a roadmap for drawing on inner strength through personal and professional challenges.

The twenty stories you are about to read are wonderful examples of the leadership, passion, and resilience my fellow refugees and immigrants have brought to their careers in Minnesota. This book offers insight into the diversity and depth of the migrant experience through the lens of those pursuing innovation in the fields of science, technology, engineering, and math. As a fellow storyteller, I am excited for the authors of *Green Card STEM Voices* as they bravely share their stories, for the thoughtful conversations this book will spark, and for the many readers who will be inspired to pursue their own dreams.

In my role as the Deputy Commissioner at the Minnesota Department of Employment and Economic Development (DEED), I see the numbers that clearly show immigrants' contributions to our state's workforce everyday. However, the stories are more striking than numbers alone. What *Green Card STEM Voices* anthology does so well is to raise awareness about the essential contributions of immigrants and refugees through their own words. These stories add a human face to this rapidly growing sector of our economy.

Additionally, the authors of these essays are paving the way for the younger generation of immigrants and refugees who are passionate about science, technology, engineering, and math. It's crucial to ensure that our immigrant and refugee youth, especially, have opportunities to grow in their careers and develop the skills needed to succeed in the in-demand professions of today and tomorrow. I look forward to the future they will help shape.

Hamse Warfa

Deputy Commissioner at the Minnesota Department of Employment and Economic Development (DEED), Social Entrepreneur, and Author of *America Here I Come: A Somali Refugee's Quest for Hope.*

Acknowledgments

Since our founding in 2013, Green Card Voices (GCV) has recorded over 400 stories of immigrants and refugees coming from 120 countries and currently living in seven US states. GCV focuses on recording and sharing digital, first-person immigrant narratives, and in 2016, we expanded our storytelling mediums to include traveling exhibits and first-person, written anthologies. *Green Card STEM Voices: Stories from Minnesota Immigrants Working in Science, Technology, Engineering, and Math* is our seventh title, following five *Green Card Youth Voices* books focused in different cities (Minneapolis, St. Paul, Fargo, Atlanta, and Madison/Milwaukee) and *Green Card Entrepreneur Voices: How-To Business Stories from Minnesota Immigrants.*

The stories collected in *Green Card STEM Voices* have brought together the intrigue that is innate in science, technology, engineering, and math with the immigrant experience. In the stories of these immigrant scientists, we are able to understand the importance of diversity in scientific drive. Because scientists often communicate through data and academic papers, the personal journeys of people in STEM are seldom heard, especially when those people are immigrants. This book thus humanizes immigrant experience in STEM by giving storytellers the chance to tell their own stories.

For all that is stated above, this book would not be possible without the twenty immigrants working in STEM from across the state of Minnesota who shared their stories and their time. They live in cities including Blaine, Bloomington, Linwood Township, Minneapolis, Maplewood, Northfield, Sartell, St. Paul, Rochester, Tonka Bay, and Woodbury. Their brilliance, grit, and perseverance inspire us all: they are the heart and soul of this work. Their willingness to share their personal stories with the world allows us to cultivate new ideas about the stories of scientists, engineers, and physicians. We are proud to share their inspiring words.

In order to select the storytellers featured in this volume, we searched our own archive—stories we've recorded over the past six years—and sought out new stories, recorded as recently as February 2019. For those stories recorded in past years, we consulted with the storytellers and recorded new material to bring their stories up-to-date. We are very honored that we were ultimately able to showcase the twenty storytellers, originating from seventeen countries, that you will read. We also sought out diversity in terms of country of origin, gender, age, race, religion, profession, and industry. We aim to share the vast breadth of experiences and demystify some of the prevalent stereotypes about immigrants in STEM.

From the inception, this project was grounded in a multi-disciplinary partnership of academia, corporate, and museum/nonprofit leaders. We would like to thank the University of St. Thomas, the Science Museum of Minnesota, and Ecolab. We are especially grateful to the amazing individuals from these organizations: Robby Callahan Schreiber, Museum Access & Equity Director at the Science Museum of Minnesota; Dr. J. Roxanne Prichard, Professor of Psychology and Neuroscience and HHMI Inclusive Excellence Director at the University of St. Thomas; and Dr. Richard Benton, IT Manager at Ecolab. These relationships were instrumental in moving this book forward.

We would like to thank our foreword author Hamse Warfa, a Somali refugee, social entrepreneur, philanthropist, and author of his own autobiography, *America Here I Come*. Hamse Warfa is currently the Deputy Commissioner at the Minnesota Department of Employment and Economic Development (DEED) where he determines the best ways that DEED can serve underrepresented communities. He is the Co-Founder and Executive Vice President of BanQu Inc., a software company working to eliminate poverty. In addition, he is the Founder and Principal of Tayo Consulting Group LLC (TAYO), a professional services firm comprised of cross-sector experts with deep experience in human development and poverty solutions.

Our many thanks also to Dr. J. Roxanne Prichard, who co-wrote the introduction for *Green Card STEM Voices*. She is a professor at the University of St. Thomas, the Scientific Director for the Center for College Sleep, and key partner on this project. Prichard's efforts led to St. Thomas's reception of the Howard Hughes Medical Institute (HHMI) grant, a one-million-dollar grant to support STEM education for all students. The grant is being used to support this education by creating a campus environment where inclusivity thrives through constant reflection, analysis, and accountability.

We express deep gratitude to Robby Callahan Schreiber, Museum Access & Equity Director at the Science Museum of Minnesota, for securing the necessary funding, hosting the book launch event, and providing the necessary institutional environment for meaningful partnership. The partnership between GCV and the Science Museum of Minnesota will go beyond the production and launch of the book, and focus on sustainable programming centered around immigrants in STEM.

We contracted with Media Active to do the video recordings of most of the stories. We especially want to thank Dominica Asberry-Lindquist, David Buchanan, and Carmela Simione, who did the videography and photography, as well as Michael Hay for his supervision. Thank you also to GCV's Public Ally/Americorps member and photographer extraordinaire, Shiney Her, who traveled to authors' places of work and took portraits in research labs, universities, and offices. This made for unique portraits that capture the work environment of our authors.

Many thanks to our crew of transcribers who spent hours listening to the audio from the interviews of the immigrant storytellers and helping us take their spoken words to the page. These individuals include Brittany Bolstad, Anna Boyer, Frances Huntley, Shana Lee, Jessie Lee-Bauder, and Renee Zhang. In addition, we would like to thank our incredible volunteers who not only helped transcribe the interviews but also met with the authors in-person or communicated via email to help them finalize their essays. These individuals include Jenni Miska, Ellen Perrault, Heidi Swanson, Jonathan Zagel, Peg Reilly, and Marta Fraboni.

We would like to extend a sincere thank you to the Green Card Voices' team. In particular, Shiney Her, Graphic Designer; Ellis Sherman, Lead Designer who created the interior of the book; and Kylee Johnson, the Graphic and Social Media Intern who assisted with the graphics. We thank Minju Kim, Video Editor, who transformed raw video footage into compelling digital narratives and designed the interior of the book. Thank you Napoor Gordon for the beautiful cover. Initially Jessie Lee-Bauder, GCV Program Associate, and later Julie Vang, GCV Program Manager did an extraordinary job of keeping the team on track. Corrine Schmaedeke, GCV's summer intern, selected quotes, while Madelyn Osmon, GCV's Fall Programs and Publishing Intern corresponded with individuals who contributed their praise and reviews of the book and assisted with the editorial work.

Beyond the above-mentioned individuals and institutions, we would like to thank the Kennesaw State University students and professors for copyediting and glossary support. Beginning in 2017, their English Department has assisted Green Card Voices by providing editorial and writing support. Dr. Lara Smith-Sitton in her role as the department's Director of Community Engagement supervised the student's final editorial work and the creation of the glossary through the incorporation of this as a service learning project in a graduate writing class. Their process included reviewing and copyediting of the essays by students in the Master of Arts in Professional Writing program, followed by a close, final edit of the entire book by Dr. Smith-Sitton. Graduate student copyediting was led by Courtney Bradford and Will Lawson. Additional editors included: Ime Atakpa, Amoshia Blakeney, Donna Cochran, Jordan Dollar, Konrad Drapela, Samantha Godwin, Jonquil Harris, Katie Metzger, Laura Metzger, Giselle Reid, and Taylor Smith. These editorial assistants were instrumental in fact-checking and organizing the glossary definitions before the editorial team and Dr. Lara Smith-Sitton finalized them.

Heartfelt appreciation goes to the book's editors: Julie Vang and Tea Rozman Clark. They worked one-on-one with the twenty authors to ensure their stories were accurately told, coordinated with diverse stakeholders and partners to ensure the high quality and sustainability of the book (and the project as a whole), and guided close to 100 people involved in the production of this anthology.

Thanks to our funders—from individuals who contributed during our 2019 Fund-a-Need at the GCV Annual Gala to generous grants from Saint Paul & Minnesota Foundation, F. R. Bigelow Foundation, and Ecolab Foundation, and funding from the University of St. Thomas, and the Science Museum of Minnesota. Many people also supported us by pre-ordering the book. Without all of you this publication would not have been possible! Thank you.

Thanks to all of our board members—Mahlet Aschenaki, Dr. Richard Benton, Debjyoti Dwivedy, Gregory Eagan IV, Andrew Gordon, Shukri Hassan, Ruben Hidalgo, George C. Maxwell, Laetitia Mizero Hellerud, Leslie Rapp, Jane Berg Reidell, Dr. Lara Smith-Sitton, Monique Thompkins and Luis Versalles—and all others who have helped our mission along the way.

Finally, and most personally, we would like to thank our partners, children, families, and friends for helping each of us put our passion to use for the betterment of society. With the above support, Green Card Voices is truly able to realize its mission to use the art of storytelling to build bridges between immigrant and non-immigrant communities by sharing first-hand immigration stories of foreign-born Americans. Our aim is to help the collective us in the US see each "wave of immigrants" as individuals with assets and strengths that make America remarkable.

Green Card Voices Team

Introduction

"Your assumptions are your windows on the world. Scrub them off every once in a while, or the light won't come in." —Alan Alda

In these pages, you will find twenty stories of Minnesota science, technology, engineering, and mathematics (STEM) professionals who immigrated to the United States. You will read about the forces that motivated these individuals to leave their homelands (safety, educational opportunity, family responsibility, love) and about the factors that drew them into a STEM career (curiosity, altruism, aptitude). You will read stories of individuals who spent years in refugee camps before finally arriving in the US with just a suitcase in hand and a few words in English, to those who arrived fluent in English and with professional STEM degrees, only to find that their diplomas carried little or no weight in the United States. They are stories about the thrill of new scientific breakthroughs and about the frustrations, failures, and false starts along the way. You will read about experiencing culture shock, discrimination in school and at work, and, of course, stories about the coldness of that first Minnesota winter.

As an organization, Green Card Voices has worked tirelessly to humanize the debates surrounding immigrants, refugees, and their families. Through video storytelling, in-person events, podcasts, traveling exhibits, and books like this one, Green Card Voices aims to build inclusive and integrated communities. Although the United States has always had immigrants, common narratives gloss over the significant impact of America's new arrivals, even though our multicultural and ethnically-diverse population has always been one of America's greatest assets.

Storytelling has been the primary way humans have communicated important values, truths, and covenants for millennia. Our minds are hungry for narratives, and we remember stories long after we forget statistics and formulas. Stories inspire, engage, and connect us across boundaries, both real and assumed. Research shows that the US-born people who are the most fearful of immigrants are also the least likely to know and interact with immigrants on a daily basis.[1] Here, we hope these twenty stories connect us as individuals, allowing us to operate with a more open and compassionate mindset. The Green Card Voices process, which typically involve asking storytellers six open-ended questions, was expanded for this book to capture the unique insight of

---

1. Pettigrew and Tropp. (2011). *When Groups Meet: The Dynamics of Intergroup Contact.* New York, NY: Psychology Press.

STEM professionals as they reflected on their scientific training as part of their lived experiences as immigrants in the United States.

According to the US Census Bureau, an estimated 43.2 million people, roughly 14% of our population, were not born in the United States. By 2050, it is expected that one in five Americans will be an immigrant. With these numbers, it should be no surprise that much of our economic, technological, and medical advancements have been due to the efforts of immigrants. In fact, immigrants are much more likely than US-born residents to work in STEM professions. In the last two decades, foreign-born Americans won almost 40% of the Nobel prizes in chemistry, medicine, and physics and won all of the United States' 2016 Nobel prizes in STEM. Immigrants are also more likely to be health care professionals than native-born citizens; 20% of pharmacists, 23% of nurses and home health aides, 25% of dentists, and almost 30% of physicians in the US are foreign-born.[2] In underserved counties, including many rural areas, the percentage of immigrant physicians rises more than one-third, thereby meeting needs addressing health care disparities in disadvantaged communities.[3]

Why are immigrants more likely to become scientists, health care professionals, and engineers than US-born individuals? One possibility is that K–12 training in math and sciences is far greater in many other countries, so immigrants arrive better prepared in mathematics and science reasoning. In the cross-national Programme for International Student Assessment, which measures reading ability, math, and science literacy, the US ranks thirty-first out of seventy countries in these competencies, and thirty-ninth in math.[4] Simply put, many immigrants may be better prepared for the intellectual demands of STEM careers than US-born individuals. Among immigrants to the United States from African countries, 40% have at least a bachelor's degree with 30% of those being in a STEM field.[5] Recruiting skilled individuals with expertise in STEM is a central feature of our immigration policies and practices.

Another possibility is that the experience of immigrating to a new country, learning new languages, and adapting to ever-changing circumstances strengthens traits like problem-solving, resilience, and determination, which are central to success in STEM fields. US-born college students are just as likely as immigrants to begin a

2. American Imimgration Council (2017, June). "Foreign-born STEM Workers in the United States." Retrieved from https://www.americanimmigrationcouncil.org/research/foreign-born-stem-workers-united-states

3. American Immigration Council (2018, January). "Foreign-Trained Doctors are Critical to Serving Many U.S. Communities." Retrieved from https://www.americanimmigrationcouncil.org/research/foreign-trained-doctors-are-critical-serving-many-us-communities

4. DeSilver, D. (2017, February). "U.S. Academic Achievement Lags That of Many Other Countries." Retrieved from https://www.pewresearch.org/fact-tank/2017/02/15/u-s-students-internationally-math-science/

5. Cownen, T. (2018, January). "Africa Is Sending Us Its Best and Brightest." Retrieved from https://www.bloomberg.com/opinion/articles/2018-01-12/africa-is-sending-us-its-best-and-brightest

degree program in STEM but are less likely to persist in one.[6] Perhaps in comparison to some of the stories you will read here—of living in refugee camps, of surviving wars, of leaving behind loved ones—the relative stress of preparing for advanced chemistry and mathematics examinations seems more manageable.

Or perhaps immigrants are motivated by the economic stability of STEM jobs. Professional organizations recognize the need to recruit and retain health care professionals who are immigrants. With our aging population and a large number of retiring health care providers, the United States is simply not training enough new health care providers to meet our growing needs. The Association of American Medical Colleges estimates there will be a shortfall of between 46,100 and 90,400 doctors by 2025, many in primary care.[7] In Minnesota, a state where almost one out of every seven residents is currently elderly, finding enough health care workers to meet the growing need for geriatric care remains a challenge. Simply put, we rely on immigrant STEM professionals to keep our population healthy.

In Minnesota, immigrants account for 8% of the state's total population, and almost 16% of the state's STEM workers are foreign-born. As in many areas of the United States, there are more STEM jobs than candidates; in 2014 there were more than fifteen STEM jobs posted for every one unemployed Minnesota STEM worker.[8] A study from the American Enterprise Institute, a non-partisan public policy research group, found that recruiting foreign-born STEM professionals benefits the local economy by creating more than double the number of jobs for US-born workers in the seven years that follow.[9]

But in the end, the real importance of this book is not about population statistics, economic growth, or even the critical need to meet our health care shortage. The importance of this book resides in the transformative power of stories. These stories are meant to inspire young immigrants who have a passion for STEM, but know no role models yet in their communities. These stories are for the educators who do not yet have examples of success to share with new generations of immigrant children. Importantly, these stories are for the Minnesota-born residents who have not yet had the experience of living and learning in different international communities. In our global economy, universities and businesses require their members to be comfortable working

6. Han, Siqi. (2016). "Staying in STEM or Changing Course: Do Natives and Immigrants Pursue the Path of Least Resistance?" *Social Science Research*. 58. https://doi.org/10.1016/j.ssresearch.2015.12.003
7. American Immigration Council (2018, January). "Foreign-Trained Doctors are Critical to Serving Many U.S. Communities." Retrieved from https://www.americanimmigrationcouncil.org/research/foreign-trained-doctors-are-critical-serving-many-us-communities
8. New American Economy (2016, August). "The Contributions of New Americans in Massachusetts." Retrieved from http://www.newamericaneconomy.org/wp-content/uploads/2017/02/nae-ma-report.pdf
9. New American Economy (2016, August). "The Contributions of New Americans in Massachusetts." Retrieved from http://www.newamericaneconomy.org/wp-content/uploads/2017/02/nae-ma-report.pdf

in diverse, multinational groups. Being able to understand, empathize, and learn from others' experiences is a necessary skill in our world, and it is our responsibility to help teach those skills. Dr. David Asai, Senior Director for Science Education at the Howard Hughes Medical Institute writes:

> "We understand that scientific excellence depends on creativity, that creativity emerges from diversity, and that the advantages of diversity are realized through inclusion…The responsibility for creating an inclusive environment lies with those who teach, mentor, manage, recruit and hire the scientific workforce, and learning the skills of inclusivity demands opportunities to make emotional connections." [10]

We at Green Card Voices hope that these personal stories help forge those critical emotional connections with readers. Often forgotten in these international narratives, but equally important to remember, are the Native Americans and the descendants of the Africans who were enslaved and brought here against their will. While immigrants often thrive in STEM education programs, US-born students of color experience additional institutional barriers to success that must be addressed. The work of GCV is to combat stereotypes and create empathy, and we hope the experience of sharing these stories extends to how we conceptualize all of those marginalized in our communities.

So, read these stories with an open mind. Allow yourself to connect with the individuals, be inspired by what they have been able to achieve, and celebrate the positive impact they have had on our Minnesota communities. Moreover, recognize that we all have a unique story to tell and that we all have the ability to inspire. Every story, and every person behind the story, contributes uniquely to the tapestry of our nation.

Dr. Tea Rozman Clark   Dr. J. Roxanne Prichard
Green Card Voices   University of St. Thomas

---

10. Asai, D. (2019, January). "To Learn Inclusion Skills, Make it Personal." Retrieved from https://www.nature.com/articles/d41586-019-00282-y

How to Use this Book

At the end of each author's essay, you will find a URL link to the author's digital narrative on Green Card Voices' website. You will also see a QR code link to that story. Below are instructions for using your mobile device to scan a QR code.

1. Open your phone camera and scan the QR code. If your phone camera cannot scan the code, use your mobile device—such as a smartphone or tablet—to visit the App Store for your network. Search the App Store for a "QR reader." You will find multiple free apps for you to download. Any one of them will work with this book.

2. Open your new QR reader app. Once the app has opened, hover the camera on your mobile device a few inches away from the QR code you want to scan. The app will capture the image of the QR code and take you to the author's profile page on the Green Card Voices website.

3. Once your web browser opens, you'll see the digital story. Press play and watch one of our inspirational stories.

STEP 1

Open up your phone camera OR download the app.

STEP 2

Scan the QR code.

STEP 3

Watch the digital story.

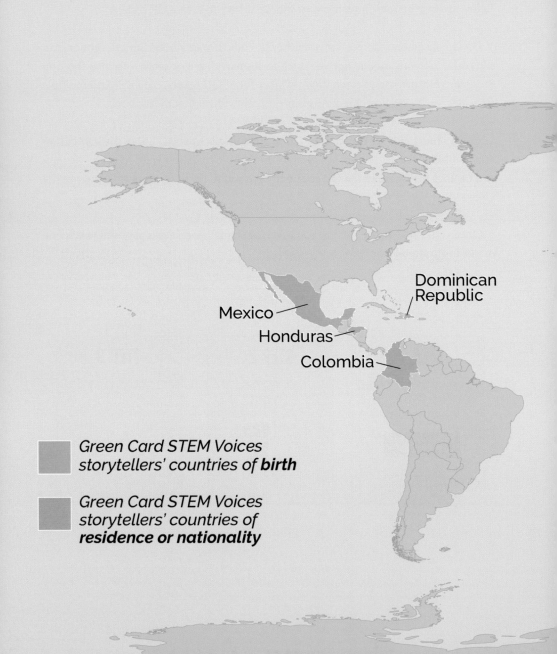

Mexico

Dominican Republic

Honduras

Colombia

Green Card STEM Voices storytellers' countries of *birth*

Green Card STEM Voices storytellers' countries of *residence or nationality*

World Map

Personal Essays

AFRICA

Jigjiga, Ethiopia

Fadumo Yusuf

From: Jigjiga, Ethiopia (Somali)
Current City: Minneapolis, MN

Field: Mechanical Engineering
& Writing

> "THERE WERE A LOT OF OPPORTUNITIES [AND] CHALLENGES. THE IDEA THAT I COULD BECOME ANYTHING I WANTED TO WAS MIND-BOGGLING TO ME. THE IDEA THAT THERE ARE SCHOLARSHIPS OUT THERE THAT ARE SPECIFICALLY TAILORED TO HELP PEOPLE THAT MIGHT NOT BE ABLE TO AFFORD SCHOOL WAS INCREDIBLE."

I was a typical child. I was going to school, took Islamic classes after school and on weekends. I grew up in a friendly, open neighborhood, and I played outside a lot. The school system there was very different from the school system in Minnesota. Schools were half-day—one month I went to school in the morning, and one month I went to school in the afternoon. I used to wake up very early, do some chores, and then go to school. On the month that I went to school in the morning, I used to come home in the afternoon, do the dishes, help make lunch, then go to my Islamic classes. In the evening when I came back home, there used to be a lot of family time with storytelling. Sometimes some of our neighbors would join us. It was a very friendly neighborhood. People often came to our home, and we went to other people's houses. Sometimes, people would just come outside on the street and play and chat. It was very vibrant and nice environment to grow up.

Moving to the United States of America

My family and I found out that we were going to the United States of America. There was an organization called IOM, and they posted the people that were approved and were going on their wall. We used to go to the wall often to see if we were posted. One day one of my family members found out that we were posted. She came home running and screaming and happy, saying, "We are leaving in nine days!" I was in eighth grade. We were supposed to leave on Monday, but I had final exams on Wednesday—I was more worried about passing my exams than I was about leaving. I told myself that I had to pass my exams. I was ready, just in case someone told us that we were not going to America. I didn't want to waste a whole school year—I didn't want to waste my education.

The whole time that my family members were going shopping and getting ready, I was studying for my exams. I went to school on Friday to take a practice exam, and then we went to the airport that following Monday. It was kind of exciting and nerve-racking at the same time. I wasn't even sure if we were going to go because the whole process took a long time. We were supposed to come back in the 1990s. The process restarted again in 2002, and we came to Minnesota in 2006, so I did not know if we would go or not, but I knew that I wanted to make sure I get my education and pass my exams.

Leaving back home, I left behind a lot of my relatives and friends. I had a lot of friends and people that had known me since I was a kid. It was kind of scary to be going to a new place, but I knew at the end of the day, my parents were making this choice for us because there were better opportunities for us in terms of education, life, and most importantly, safety.

When we left, we had to go to Addis Ababa, the capital of Ethiopia. We then went to Bole International Airport, then to Amsterdam. I remember we had this huge IOM bag, and because nobody in my family spoke English, there were people to guide us at every airport that looked for the IOM bags. We sat at Amsterdam for six hours, and it felt like a really, really long time. Finally we came to New York, but it was a little bit confusing because it was supposed to be night, but when we landed, the pilot said, "Good morning." I was so confused and thought, "How is it a morning?" It was supposed to be night, but we landed at 7:00 a.m., but we didn't get to Minneapolis until midnight. I remember we sat at the airport, and we didn't really know what to do. Someone guided us to chairs and said, "Sit here," and we just sat there for hours. I walked up to one lady and pointed to my wrist and said, "Time?" I was so confused. Is it a morning? Is it a night? What time is it? I had to get to the bottom of it—that is how I found out it was 7:00 a.m. But the thing that struck me the most was that the flight attendants were smiling the whole time. Where I grew up, you don't really smile as much, especially to strangers. It was kind of weird—I wondered why they were smiling.

We came in June. I thought it was going to be cold and snowy, but it was a hundred degrees outside. I was like, "It can't be a hundred degrees. We would be cooking!" I remember learning in science class that water boils at one hundred degrees. I was like, "We would have all been dead by now." And she said, "No, it is a hundred degrees." And I was like, "No way, it's not possible." So she turned on the TV and was like, "See? The TV." And I was like, "Your TV's probably lying." And her husband was sitting there just laughing because he knew we were on two different systems: Celsius and Fahrenheit. The whole English system was like the biggest culture shock to me.

The food was great. I was actually really happy with food. I ate so much pizza when we first came. It was the first time that I had eaten pizza. I remember seeing it in papers and sometimes on commercials on my neighbor's TV. It was amazing, and every day I asked my aunt, "Can we get pizza?" She'd say, "We can't have that every day!" When we came, my first question was "When can I start school?" And they said, "School is out. It is summer, so there's no school." And I was like, "But I want to go to school. I need to get registered." I then realized that there's different times for school.

I did not know much English. I knew very few words, like water and time—basic things. I was put in the lowest ELL class; I had to learn the language and all the other subjects too. I was trying to navigate through a school system that was drastically different from what I was used to. I couldn't really communicate with my teachers as much as I did back in Ethiopia. I had to learn the culture and the language and figure out how this education system worked. I knew I wanted to go to college. I had to figure out how I was going to go to college and what do I need to do. I was just talking to any teacher that I could find that I could communicate with. The biggest challenge was really knowing what the opportunities are, how to learn enough English to get out of ESL, and how to get into college. I also had to find out that there's something called the ACT that I needed to study for and take. I was in this completely new environment—my parents never went through this system. My family members had no idea. I had to figure out how I was going to succeed and how I was going to do all these requirements that I had to meet.

STEM Education and Career

From a very young age my mom taught me that it was okay to be myself. It was okay to be curious, it was okay to have dreams, and it was okay to be a little weird. For example: One day, I got two rocks and a charcoal. I wanted to see how many small pieces I could break the charcoal into. My mom saw me sit outside in the sun, for what seemed like hours, as I hit the charcoal with the rock. She asked me what I was doing. When I told her what I was up to, she looked at me and said, "Let me know what you find out." And left me alone. To me, this was my mom's way of teaching me to embrace all aspects of my personality—including my relentless curiosity and weirdness.

I kind of had a personality for STEM—I was very curious, liked learning different things and loved experimenting. I loved trying to make new things. I also liked eating foods that I did not know about. I remember every time I travelled to a different city with my mom, when we went to a restaurant I would ask for the food that I didn't know about. My mom made a deal with me that even if it was a monster with eyes that I would have to eat it because I was wasting a lot of food if it wasn't that great.

When I was going to high school, I had a teacher called Dr. Claire Hypolite. She has a bachelor's and a PhD in chemical engineering, but she specifically chose to teach at that high school because she felt that's where she could make a difference. One day she came to me, and she said, "Hey Fadumo, what do you want to be?" And I was like, "Oh, I'm going to become a doctor because that's kind of the only thing that I have ever wanted." I knew I wanted to do health studies and make a difference. And she said, "Oh, okay. Are you sure about this? Have you looked into other career options?" And I was like, "No, I don't need to. I volunteer at children's hospital. I'm going to become a doctor." Then one day she asked me, "Hey, I'm starting this after-school program called 'Invention Club.' You can come and invent whatever you want, and I will give you resources." I was like, "Oh, I don't have someone to take me home—I don't have a ride." My parents didn't drive, and if I missed the school bus, it was really difficult for me find transportation back home. She said, "I'll give you tokens, and you can use that to go home." I was like, "Great!"

It was my job to cook at home, and I would rather read and do school-related stuff. I just felt I was wasting a lot of time, and I wanted to invent a machine that I can put all the ingredients in and press a button, and in five minutes it would give me the food that I wanted to make. After looking into it a little bit, I realized I didn't really have the skills to do that. But then, I read an article about women in remote areas that were infected by HIV who might not have access to devices that can tell them that they have an HIV. And so I wanted to make something, so I told Dr. Hypolite, "I want to invent an HIV testing device." She said, "Cool, how are you going to do that?" I didn't know, so I started looking into it and trying to design it. I would think, "Okay, how do people figure out if somebody has a disease?" And reading what methodologies are used, she introduced me to the field of microfluidics where you can manipulate how fluids flow and stuff like that. She took me to different labs and showed me how some people were doing research and how they were using those devices. Essentially my idea was to make a device where your blood can flow through series of channels, and as it flows through, different things happen that can tell you whether you have HIV or not.

When the school closed, I would go to school over the summer, and she would sometimes meet me downtown at Central Library. She taught me how to do research, how to read research papers, how to write research papers in MLA format, and stuff like that. By the end of that project, I was like, "Who makes these things? I want to do this." And she was like, "Great, I knew you would love this! Engineers!" And I was like, "Oh, engineers. I thought they built roads and outside in the sun all day." She told me, "No, there's a whole other side of engineering that people don't always know." She introduced me to the world of the medical device industry, and I was like, "Okay, where do they work?" She told me one specific company. When I got my bachelor's in mechanical engineering, my first job out of college was actually the company she told me that day. While working there, I got accepted to master's degree program for mechanical engineering, so I sent her an email. I was like, "Guess what? I've become an engineer, and this is where I work—at the company you told me about." I just sent her a thank you email. Dr. Hypolite opened my mind to the world of STEM, particularly to the world of medical devices. She opened my eyes to the endless possibilities in this world, and to the many different ways that I could make a difference and contribute to humanity. And that's how I ended up in STEM.

Challenges and Opportunities

There were a lot of opportunities, and there were also a lot of challenges. The idea that I could become anything I wanted to was mind-boggling to me. The idea that there are scholarships out there that are specifically tailored to help people that might not be able to afford school was incredible. Paying for school was my biggest worry because I couldn't take loans because of the interest. So I applied to a lot of scholarships. I think I applied to over ten or twelve scholarships, and I ended up getting about seven of them. The biggest one was the Gates Millennium Scholarship. The Gates Scholarship was just the greatest opportunity for me; it really enabled me to not only get my bachelor's but also my master's. I couldn't have asked for a better place to be in the world. Not only could I study everything that I wanted, but there were people who were nice and kind enough that were willing to give me their money—the money that they have earned, that they have worked hard for, so I can have a better future and get an education. To me, that was really amazing and very inspiring.

I had some basic challenges. I was first generation and didn't have a lot of guidance, so it was a lot of having to figure things out on my own. I had to figure out the opportunities, requirements, how to meet people, and even that there is something called office hours that you're supposed to go to. Going to the University of Minnesota as a PSEO student helped. I was going to the U of M when I was still in high school. There were some challenges, but at the end of the day the opportunities greatly outweigh any challenges that I had.

My first lesson is to be comfortable with who you are and your identity and in your own skin. Because when you're in an environment and you're very different from everyone else, sometimes it can be a little bit intimidating, and sometimes you're like, "Oh my God, these people have been prepared for this since they were kids." Sometimes you feel kind of out of place, but just be comfortable with who you are and with your identity. Use your identity as an asset, rather than seeing it as a liability or something that can hold you back or diminish you. And regardless of where you are in your life or what you know, your culture, or your community, everybody has a challenge—even the people who were born here, people who just came, and the ones who've been here for generations. Every human has their own personal challenges and struggles they're going through, so just keep that in mind. You're not the only one struggling or going through some stuff. That is really important and kind of is what kept me going, just being comfortable with being the odd one out.

My second lesson is just show up. Some days you feel like a failure. You will experience some setbacks, but that should never be the end of your goals and your dreams. It should be like, "Okay, I had this setback, or a little failure here, but it's okay." Bounce back and ask yourself, "What can I do different? Are there any resources or help that I can get so I don't have that setback anymore or so I can lessen the impact that it had on me?" Those are kind of the key things that I think helped me and still help me in the industry that I am in.

Contributions and Impact

I work in the medical device industry as a product/process development engineer, which is an interface between the research and development and the manufacturing of medical devices. I work with a lot of people, taking concepts and turning them into actual technologies that can help people. I support getting new technologies to the market by supporting clinical builds to get the device approved by regulatory bodies like the FDA. My contribution is to get new technologies and devices that can help people into the market. It is helping with the design of the new technologies . . . taking a design or concept and figuring out how to make that technology . . . what materials to make it from. I also sometimes test devices and write reports required by regulatory bodies. That is kind of what I do day-to-day at work.

In terms of my community, I serve as a board member at a local STEM school. I go to schools to talk to kids about my career path. I show them some of the opportunities and available resources to learn or grow in my industry. I also show that you can still be who you are, have many different aspects of your identity, and have a career and succeed. I share my experiences to show you can overcome any struggles. I mentor younger students in my community. One of my mentees is actually graduating this year from U of M in mechanical engineering. I am so proud of her!

Personally, I love writing: I write Somali poems and short stories. I am working on publishing my first novel. I share my writing through my blog, fadumoyusuf.com. I love taking quiet nature walks in the summer and reflecting. On weekends, I go to classes to learn Islamic Sciences to better understand my own faith. I also give back to my community through my writing. I wrote a short story for the Children's Theatre called "The Strawberry in a Banana Farm." They have a program where they go to schools and teach kids about different cultures and acceptance through stories, plays, and stuff like that. They used two stories that I wrote for them. The idea of one of the stories was basically to be comfortable with who you are—be comfortable with your identity. It is okay to be different—no two people on planet earth are the same. Embrace your identity, the depth of who you are, and all the different aspects of your life. Those are some of the small contributions that I have made so far, but I do plan to contribute a lot more . . . I am just starting.

Appreciate the heroes in your life—those who uplift you, who encourage you, who show up for you, and who help you find and amplify your voice. Even more importantly, be a hero for someone else. Amplify, uplift, and help them so that all voices rise.

greencardvoices.org/speakers/fadumo-yusuf

Phra Phutthabat district
of Saraburi Province,
Thailand

ASIA

Thai Chang

From: Phra Phutthabat district of
Saraburi Province, Thailand (Hmong)
Current City: Maplewood, MN

Field: STEM Education

> "THE MORNING AIR WAS SO FRESH; IT WAS ACTUALLY AMAZING. THAT'S ONE OF
> THE REASONS I NOW LIKE CAMPING AND SLEEPING IN A TENT. THE MORNING AIR
> AT A CAMPSITE REMINDS ME OF MY FIRST MORNING HERE IN THE UNITED STATES."

I grew up in Wat Tham Krabok, a refugee camp in Thailand. Wat Tham Krabok is a Buddhist temple, and in the 1990s, the monks offered part of their land for the Hmong refugees as shelter. The camp was surrounded by mountains but later gated by barbed wire. From a child's perspective, it was an exciting and adventurous place to be—the mountains and the temple were our playground. With a group of friends and slingshots in our hands, we hiked up the mountain to explore the woods and hunt birds for a meal. As little as we were, conquering the mountains was effortless.

Beside the mountains, we also loved to go play in the temple and pick fallen fruit from the trees. Walking around, playing tag, and swimming were regular activities for us, and if you got hungry, fruit trees were everywhere, especially mango trees. Picking fruits from trees was forbidden; however, if any fruit fell on the ground, it was free for anyone to eat. We loved going to the temple during a rainstorm. The storm would shake mangoes off the trees, and we would battle our way through the storm to find the fallen fruits. Another way of getting fruits from trees were to help the monks with their daily chores. The mountain and the temple were places to be for the curious kids. As a kid, the camp was full of exciting experiences and endless adventures. However, looking back as an adult, I realized that there were no realistic nor legal ways to get a higher education or a job. A path to citizenship was never possible nor affordable. We were confined and not allowed to be outside of the camp. Many families struggled to put food on the table. There was no opportunity for one to explore their hopes or dreams.

My mom worked really hard to provide for my family. She made traditional Hmong clothes that Hmong people wear to celebrate the New Year. She would then send them to my sister in the United States to sell. My sister was here before us in the 1990s. She moved to the United States after she got married. When my sister was able to sell the clothes, she would send money back to support our family in the camp. I was fortunate that my mom was able to send me and my brothers to Thai school. Most of the people could not afford to go to school.

I remember when I was twelve years old, I snuck out with my friends and slept over at their place to try to work on the field. Around midnight, we went to a "tshav npas" (soccer field), where Thai farmers come pick up people for labor work. The field was filled with people of all ages, anywhere from ten years old to over sixty years old, all eagerly and desperately want to work that day. It was complete chaos. It's a mile stretch from the road to the tshav npas. Right when we could see the headlights of a truck approach from afar, people would start running towards it with no safety in mind. I recall a friend screaming, "We have to get on that truck! We have to get on that truck!" By the time the truck reached the tshav npas and came to a complete stop, it was already filled with people. Whoever was on the truck got to go work on a field that day. Those that did not get on a truck before sunrise would have to go home with their packed lunches and without work because farmers needed workers on their field before sunrise. My friends and I were able to get on a truck, and we went to a bean farm. We picked beans all day, and I got paid sixty baht, which was about one-dollar-and-fifty-cents back then. It was my first and only day working in the field. I was not able to gather as many as my friends and was paid much less. It was both an amazing and exhausting experience—I did not know how people could keep doing that every day.

Moving to the United States of America

The Hmong people in the camp were told about immigration two years prior to the move. I started seeing people go through interviews and check-ups for approval to move to other countries. This process felt extremely surreal. There were rumors in the past that made many people doubt that it would be possible to come to the United States. The entire perimeter of the camp was gated with barbwires to keep the refugees in and others out. The lack of communication from the authorities made people in the camp panic. It felt like years and years of waiting back then.

In spring 2004, people were being relocated to different parts of the world. The United States was one of the options that we could choose from; Australia and France and a few other countries were also accepting immigrants. I remember going to the hospital where they do all the interviews and checkups and seeing the map of the world on the wall. I didn't know where things were, so friends pointed out the United States on the map. They were saying that the route we will be flying is going to be from Bangkok, to Japan, to the United States. But on the map, the United States was on the opposite side of Japan. As a curious kid I said, "Why are we going backwards to Japan and then flying behind the map to the United States?" I didn't realize that the world is a sphere and it was shorter the other way around. So I was thinking, "Why would we go all the way to the back of the map to get there?"

I was thirteen years old when I came to the United States. My family was part of the second group that was coming here and there were about four families, including mine. We packed up everything and decided what to take and what we wanted to leave behind. It was especially tough for my mom. She was the one who made everything happen for life in the camp. She was most attached to her sewing machine, but it had a large wooden base, so we decided to leave it behind because we couldn't carry it on the plane. We didn't have much information about what would be available in the US, so it made it hard to make decisions about what to bring. We mostly packed our clothes and little belongings that we could put in our backpacks or suitcases.

Getting on the bus in the camp to go to the airport was the moment that I realized that it's actually happening—it was so surreal. I physically was going to leave the camp and everything else that I knew. I was tearing up as the bus was leaving the camp. There were lots of people that surrounded the bus, and as it moved, a lot of people were crying as they waved goodbye to us and to their families. It was horrifying that none of us knew what to do or where to go after we stepped into the airport in Bangkok. I am amazed how, at each airport, they arranged a guide to help us transition to the right spot and to the next plane. Bangkok was not as challenging, but right after that it was like walking in blindness—you didn't know where to go and what to do . . . you were hearing things that didn't sound familiar. We didn't really know how to communicate—questions such as "Where's the bathroom?" or "Where do I go next?" were not possible. That's when we realized that it's going to be a challenge from this step forward.

Physically, it was exhausting. It was everyone's first time on a plane. It was a long transition from the camp to Bangkok. Most of us were not used to being enclosed in a car, so the motion sickness was unbearable. I don't recall much in terms of where or what stops we had besides the one in Japan and landing in Minneapolis. We chose to come to Minnesota because my sister lived here. She was there at the airport waiting for us to arrive. When we landed, it was June 24th around 9:00 p.m. It was in the middle of summer, and we could still see the sun in the horizon as we drove to my sister's house. I remember this clearly because in the camp the sun sets much earlier, around 6:00 p.m. Along the way I saw how different it was compared to the camp. We were used to dirt roads where every step would stir the dust from the ground. Here, you see green grass, pavement and sidewalks, and barely any dirt. We were also used to seeing huts and houses built out of wood and bamboo and walls built out of cardboard, or whatever you can cover the side of your house with. It seemed like everything here was really organized and planned out.

We stayed at my sister's place for the first month. I remember my first morning at my sister's house—the morning air was so fresh . . . it was amazing. That's one of the reasons I now like camping and sleeping in a tent. The morning air at a campsite reminds me of my first morning here in the United States. That first week included going to the doctor a lot, getting shots, and getting blood drawn for testing. It was really intimidating, but the biggest challenge was getting there in a car: I was still not used to sitting in a moving enclosed car, and I had motion sickness for months—it was the worst feeling in the world. One thing that I realized is that you have to jump in your car to go anywhere; everything was no longer within walking distance. My sister and brother-in-law took us to visit relatives, and also lots of them came to visit us for the first time. They were excited to see us here.

I remember meeting the neighbor for the first time who was not Hmong. I tagged along with my niece and nephew as they hung out with their American friends. I was amazed with the way they could communicate with the language that I was unfamiliar with. I remember saying to myself, "How am I going to learn how to speak the language?" I couldn't pick a word out of the sentence or even out of the conversation. I couldn't tell where the sentence started or stopped. I worried about starting school.

Learning English and STEM

I went to LEAP High School where a lot of international students go to learn how to speak English. My first time going to school and seeing people from the camp was a relief—I knew some people. I met a lot of my friends from the camp and got comfortable faster than I thought. We spoke in Hmong to each other. Then I realized that if I am going to live here, I needed to learn to speak English. That's when I decided to go to a different high school so I could expose myself to people constantly speaking English. I decided to go to Harding High School. It was really different walking into Harding seeing people shaking hands, hugging, and kissing each other in the hallway. People in the camp and at LEAP didn't really show that kind of affection in public. It seemed like everyone knew everyone else, except me. I felt alone. I recall holding my schedule in my hands, and I didn't know where to go or who to ask for help. Pushing myself to ask for help was a huge step in getting to know others. My sophomore year I joined sports with the hope to get to know more people. Through sports it became easier to get to know others and make friends. From there on making friends didn't seem like much of a challenge.

I have always loved robotics. I remember as a kid, even back in the camp, I used to take out broken cassette players and take out the motors and connect them to a battery and watch them spin. I did not know how it worked or why it worked, but this curiosity sparked my interest. When I was in high school, I was interested in making stuff around the house. I loved watching MythBusters, seeing robotics and stuff being tested. It was enjoyable to watch and see how things work or not work. To me, making stuff was fueling a passion, and it was something I wanted to go to school for. Putting things together and seeing how things work tends to come easier to me than language and literature. Therefore, I ended up enjoying and excelling in science and math more than other subjects.

In 2008, when I was a junior in high school, Oanh Vue, a crew manager from the Kitty Anderson Youth Science Center, came to my school to recruit youth to join and work in the museum. I was appreciative that she came to my school and talked with us directly rather than put out flyers because of the language barrier at the time; it was hard for me to just go out and look for a job. She was recruiting for an invention crew where you get together and learn how to solve problems through making and inventing solutions to solve challenges or tasks. It was something that aligned with what I love to do. I tried to encourage a couple of my friends to apply with me, but they had different interests and did not apply. It was something that I knew I wanted to do, so I stepped outside of my comfort zone and decided to apply for it. It was my first interview ever. I wore shorts and a t-shirt and met with the crew managers Dan Haeg and Oanh Vu. It was great being able to walk through a museum and see all the cool stuff that they do that inspires others—I wanted to be a part of that. After the interview, Dan called me and said that I was on the team. I really enjoyed the challenge and work and appreciated the opportunity that allowed me to explore my interests.

I still remember the first challenge we did. We broke up into smaller groups, and each did different parts of the project. We programmed a cricket board to make a booby trap. We made a laser security system—a laser pointing at a receiver across the hallway that when someone walked past it and blocked, the laser from the receiver made an alarm noise. I finished with the program when I graduated high school as it was only for high school students. It was an amazing experience which continues to influence me to create opportunities for other youth.

Career in STEM Education

After high school, I went to the University of Minnesota Twin Cities (UMN). When I was there, I was undecided on my major. I wanted to pursue electrical or mechanical engineering; however, getting into the College of Science and Engineering was out of my reach. Language barriers were still a problem. I decided to major in Youth Studies instead with the intention of influencing and creating opportunities for young people to explore Science, Technology, Engineering, and Math (STEM) at an early age. This way, when they get to where I was, it wouldn't be as big of an obstacle. Also, I wanted to do what I love, which is making stuff, as a hobby instead of a career. I graduated with a Youth Studies major. In my second year at the UMN, I applied to the Science Museum of Minnesota to be a Maker Corps Intern for the summer. It's an initiative to recruit interns for different hubs around the country—the Science Museum of Minnesota was one of the hubs. We developed hands-on activities for the Activate program. Every Saturday, tables were set throughout the museum floors with different science, technology, engineering, art, and math activities for visitors of all ages to experiment with. The program name was later changed to Play, Tinker, Make, and Engineer. I was one of the program coordinators for approximately three years and a Learning Technologies Specialist for two years. I have learned a great amount about leadership through the program by leading workshops for volunteers, teachers, and librarians.

After my position as the Learning Technologies Specialist, I joined the Exhibit Developer Team. As an Exhibit Developer, I worked with the National Informal STEM Education Network (NISE Net) to develop hands on science activities kit: "NISE Net is a community of informal educators and scientists dedicated to supporting learning about science, technology, engineering, and math (STEM) across the United States." After a year of being an Exhibit Developer, an Exhibit Fabricator opportunity opened, and I applied. This is my current position. I was thrilled to be able to join the Exhibit Fabricator Team. I love the challenge of making and turning an idea into exhibit components that inspire learning. It is rewarding to think that the work I do activates and excites people about STEM.

I used a lot of "I" throughout the story, but all of my works were done collectively and with many amazing teams at the Science Museum of Minnesota. I would not be where I am without the help and support from my coworkers. The museum is a big institution; however, I feel like it is a small community where everyone knows everyone. I love working at the museum because of the people and the mission it stands for: "Turn on the Science: Inspire Learning. Inform Policy. Improve Lives." Even though it has been over a decade, it is still unreal to think of where I was and where I am now. The refugee camp was a great place to run around, but it was a place that lacked hopes and dreams. I am grateful to be where I am and thankful for those who helped me along the way . . . you know who you are.

greencardvoices.org/speakers/thai-chang

ASIA

Gujranwala, Pakistan

Aasma Shaukat

From: Gujranwala, Pakistan   **Field:** Medicine
Current City: Minneapolis, MN

> "I DIDN'T KNOW THE WORD GRIT WHEN I FIRST STARTED BUT NOW I COMPLETELY UNDERSTAND WHAT IT MEANS . . . IT'S PERSEVERANCE IN THE FACE OF FAILURE OR REJECTION AND WHEN ONE DOOR CLOSES, YOU HAVE TO KEEP KNOCKING UNTIL MORE DOORS OPEN."

I was born in Gujranwala, a small city near Lahore in Punjab, Pakistan, which is the middle of the country. I grew up mostly in Lahore, the cultural heartland of Pakistan. I belonged to an upper-middle class family. My father was in the civil service, and my mother was a schoolteacher for a while but mostly stayed home with us. I have two younger siblings. My childhood was very happy, surrounded by family: a lot of cousins, aunts, uncles, and grandparents. It was wonderful to grow up in that environment.

I was inspired to go into medicine by my grandfather who was a physician. I heard all these wonderful stories about him, which may not have been true but were inspiring nonetheless. The stories were that he was the only doctor in the village and that he could fix everything from severed heads to people with nerve disease. That was very inspiring. I had a natural affinity towards math and science, so I decided to pursue medical school. Towards the end of middle school, I started to think about what I would do next.

Up through medical school, the education in Pakistan is just absolutely wonderful. It's top class, and I was fortunate enough to compete for some of the best medical schools and be granted admission into them. But then beyond medical school, when I saw what I would be doing next, I wasn't very clear if Pakistan was the right place for me. It still is a developing country, and the opportunities for postgraduate education are not that great. I also wanted to do research.

When I started thinking about research opportunities, I really had to think of going beyond Pakistan and thinking of places where research opportunities were abundant. That's when I decided I would perhaps pursue another master's degree in biostatistics and epidemiology to learn the research skills and perhaps be in a place where there are more research opportunities.

My other inspiration was Mother Teresa, who we saw growing up on television. She did fabulous work in India and Pakistan, and I wanted to do something similar where I would reach out to the public instead of just sitting in an office waiting for patients to come to me.

Moving to the United States of America

I thought about different countries around the world, and really the United States stands out because the opportunities here are unparalleled. There is a lot more funding, and there is a lot more infrastructure, so I gravitated towards the United States.

Now, looking back, I realize that there's a lot that I left behind, but I didn't realize it at the time. At that time I was in my twenties, full of energy and excitement, and I felt like I could conquer the world. I remember it being a very active decision that I would go to the United States. I applied for a student visa. I took lots of exams to get into top-notch schools. I remember being very excited to be leaving, but I thought it was temporary. I thought after I did my education and completed everything I wanted to do, the whole purpose was to come back to Pakistan and serve the people, but things changed over the years. But at the time I was leaving, I didn't know that.

I literally came with a suitcase and a few photographs of things I wanted to keep close. I remember saying goodbye to my family, but I said, "I'll be back very shortly." Now looking back on it, what I missed out on is my younger brother who was at that time eight years old. I missed out on his childhood because I never went back. Both my grandmothers who I was very attached to passed—one of my grandmothers actually lived with us, and she and I were roommates. I never got to say goodbye to her—she passed a few years later, and I couldn't go back. Those are the things I left behind. Now when I look back, it seems there's a lot that I left behind.

I also left behind a whole family and a culture. I left behind an identity. I knew how to dress; I knew where to go and who my friends were. I was very comfortable in that life, and I left all that. It didn't sink in at that time, but over the years, when I look back, it seemed like a pretty large cost. I still remember my journey from Pakistan to the US. I went from Lahore in Pakistan and first arrived in Orange, New Jersey, where I had some family. From there I made my way to Baltimore, Maryland, because that's where I was admitted to pursue a master's degree in the School of Public Health at Johns Hopkins and then essentially stayed in Baltimore for the next few years as I completed a master's degree and did some postdoctoral work.

Baltimore was my first home in the US, and it was the first time I experienced renting an apartment and understanding how to open a bank account and how a debit card works because those things were all completely new to me. I still remember that it was February and the air was very crisp. I noticed how clean everything looked: the trees, the plants, the buildings, the roads. Everything was clean and neat and orderly compared to what I have been used to.

I really enjoyed how easy it was to go get food. Fast food as a student was a big convenience. It was very exciting to find my own apartment, figure out where I was going to live, figure out how to navigate the bus system—how to get back and forth from campus. I also had to determine how I wanted to dress, and how I was going to interact in my new environment. What was challenging was not having done any of these things before. Not having a credit history in this country and not having immediate family members available for help if I fell ill as well as having nobody to fall back upon—those were some of the challenging things that I remember in the first few months.

STEM Career and Life in Minnesota

What inspired me about STEM was my passion for math and science. My father told me, "You should always try to train in a field where you have some skills that are valuable and you're not dependent on, say, a government job or something else." When I was trying to pick between science and math as a woman coming from my country, I had no idea what math would look like, so I gravitated towards medicine and science. I thought I would integrate being a researcher into my work and take on all of Pakistan's medicine problems at a public health level. That's where I focused a lot of my energy, so I decided to go into medicine. My other passion is education, and I really wanted to be able to inspire the next generation and teach medical students and trainees at different stages as they progress and become scientists and researchers.

After completing my postdoctoral studies at Johns Hopkins, I pursued medicine with the idea to go to the best institution possible. The barrier that faced was that I was not a US citizen. The kind of visa that I was on kind of limited me as to which places I could go and what exactly I could do there. For instance, I couldn't really do much research because of the special kind of visa I was on. From Johns Hopkins, I went to the State University of New York in Buffalo, New York, where I focused on internal medicine. Then from there I went to Emory University in Atlanta, Georgia, and did advanced training in gastroenterology. My initial research training was at Emory. From there I was looking for opportunities where I would be able to combine the newly-learned skills I had, be able to practice in a teaching setting, do research, be a clinician, and see patients in a clinic. But, I also had to work within the restrictions of my visa. That's what brought me to the University of Minnesota, and I've been here ever since.

The opportunities have been wonderful along the way. I would say I've been able to knock on doors and find people willing to give me opportunities. I've literally knocked on people's doors and said, "Do you have a research database that I could use. . . . I have this research question. . . . Can you point me towards resources of how I can go about answering it. . . . " And everybody has been very, very supportive by-and-large. A challenge, of course, was that I was not a US citizen so I could not apply for federal grants—federal grants are somewhat restricted to US citizens. In order to achieve my goals for research, I often had to work with others and be creative in how I could navigate my goals within the resources I could access and utilize.

I've found Minnesota to be very, very responsive and open to immigrants and new ideas. This was also the only place where people would ask me, "Where are you from?" It's a pretty typical question when they first meet you. I would say, "Pakistan," and they would say, "What city in Pakistan?" I found this to be the only place where I had lived where people would ask me this. I was surprised and would respond with, "So you know cities in Pakistan?" People here pride themselves for being very educated and very literate, and I felt assimilating here was not difficult at all. There were plenty of opportunities, and people were all so very willing to help when you asked them.

What restricted me I feel is that initially I was held back by my own fears and insecurities. There is a lesson in this. What I've found is if you reach out, most of the time you get the answers that you are looking for. There's a lot of help to be had, but you have to ask. The other thing is also not being discouraged by rejection. There is plenty of rejection. I didn't know the word grit when I first started, but now I completely understand what it means. Essentially, it's perseverance in the face of failure or rejection, and when one door closes, you have to keep knocking until more doors open. That's the only way to go forward. I've found people are very supportive of your efforts, so don't try to do this alone. Please reach out.

I've made a wonderful life in Minnesota. I've been very fortunate. My husband is also a physician. We have three young children ages seven, five, and two. So, it's a busy household. I have taken up a lot of Minnesotan sports and activities. I've learned cross country skiing. I enjoy running and biking, and my children play hockey. I would say we've pretty much assimilated in Minnesota. I also have a closet full of very warm jackets and layers, so I am prepared for just about anything.

Contributions and Impact

I work full-time, and my work is divided into three parts: about one-third time as a clinician, one-third being a researcher, and one-third being an educator in the medical school and at the University. I find all those roles very, very fulfilling for different reasons. I am researching colon cancer and trying to understand how we can prevent colon cancer through diet and lifestyle, as well as what predisposes people to cancer and if one could modify some of the risk factors for it. I am also studying the role of stool transplant in curing certain diseases. I find that work very fulfilling.

I have many grants now, including federal grants—almost about ten million dollars' worth of grants and a whole research lab and staff that works very closely with me. It's very gratifying work, and at home I still get to do all the cooking with my family. We enjoy being active and being outdoors; we enjoy walking and biking. We experiment cooking all kinds of wonderful foods: the cuisine that I brought with me and the things that I've learned in Minnesota. I have a recipe for a tater tot hotdish now. Those things are what really endear Minnesota to me.

I feel we can contribute in meaningful ways wherever we are. One way I feel I do this is through research. My research is aimed at understanding how colon cancer develops and creating strategies to prevent colon cancer. It is the third-most common cancer in both men and women and the source of a large number of premature deaths. So, learning how to reduce that has been my goal and improving screening at the population level is where my research is focused. I feel I've made contributions by adding to the literature, which has led to changes in guidelines and how we treat people and how we provide expert opinion. I speak at national and international meetings to disseminate the information we have in order to move this field of research and study forward together, with others. Finally, being a role model for young women in STEM, in particular, by encouraging them to follow their passions and not to be discouraged by rejections is also a meaningful contribution that I try to make.

SCAN TO WATCH VIDEO

greencardvoices.org/speakers/aasma-shaukat

AFRICA

● Nakuru, Kenya

Getiria Onsongo

From: Nakuru, Kenya
Current City: St. Paul, MN

Field: Computer Science

> "GROWING UP, THINGS LIKE RETIREMENT WERE NOT THINGS WE TALKED ABOUT BECAUSE YOU WERE WORRIED ABOUT EATING FOOD THE FOLLOWING DAY OR WORRIED ABOUT SCHOOL FEES. NOW IF YOU DON'T HAVE SAVINGS FOR SIX MONTHS, PEOPLE LOOK AT YOU LIKE YOU'RE IRRESPONSIBLE."

I'm the second born in a family of six kids. I have four brothers and one sister. I grew up in Kenya, and even on Kenyan standards it was an above-average livelihood. I'd say my life there was pretty okay growing up with parents who were slightly better off than average in Kenya. My mom was a primary school teacher, and my dad worked for the government in the agricultural office. I went to one of the better schools in my area and did okay there. Then I went to Nakuru High School where I did my ninth to twelfth grades. I finished high school, and I passed well enough to go to the local university. So, I applied.

My uncle saw a scholarship advertisement in a newspaper, and he sent it to me and said, "You know, instead of sitting out for one year before you go to university." I think he was more afraid that I'd get into bad crowds and start doing bad things, and this was a way for me to keep myself busy. I applied for that scholarship to what I thought was a university in India. It actually ended up being eleventh and twelfth grade, but I went anyway. I went for two years. It was a bit contentious in the family because my dad thought that it was a scam and that people were just trying to steal money from us. Also, I'd already been admitted to a university in Kenya, Nairobi University, to do civil engineering.

I got the scholarship to go to India—I was fortunate to get it. There was a small application fee, and my dad did not want me to apply for the scholarship. Fortunately for me, Louise Leakey, a well-known anthropologist, was part of the scholarship committee, so I convinced my dad it could not be a scam. I asked him to let me apply, and he let me. I got called for the interview. I went there, and I did well enough to get the scholarship to India.

Moving to the United States of America

The plan was to stay in India for two years and come back to Kenya to attend university, but then when I was in India, most of my friends were applying to American and UK universities. I also applied to a number of universities in the UK and the US, and I got admitted to a few. I chose Macalester College because they gave me the best financial aid package, and that's how I ended up in the US.

I'd say my journey started when I went to India. If I hadn't gone to India, I don't think I would be in the US. I was in India for two years; then from there I came to Minnesota. I've been in Minnesota since. I was in Washington state for about a year in between my master's and my PhD degrees, but then I came back to Minnesota. My experience in a new country as a first-year student seemed natural and normal—what anyone would go through if you go to a different country. But in looking back, I think I had a lot more challenges than I thought I did at the time. I think there was a lot of excitement of being in the US, the land of opportunity. For one, people always have a hard time listening to my accent and to what I'm saying. My intonation is slightly different, I later found out. For example, I recently got engaged, and my fiancée told me that for the first two or three weeks we hung out together, she did not understand half the things that I said—she just used to smile. I guess it's a good thing she found me attractive at the time.

Also, it seems that for people who have a common culture it's easier to start a conversation. If you grew up in the US, maybe you like football or maybe you watch The Simpsons or maybe you know what is going on the most popular TV show at the time. I remember when I was in my first week, just having a conversation with anyone who had not traveled from outside the country was difficult because we didn't have much in common. Later on, after I started taking classes, we could talk about things that we had in common, which was mostly homework and things like that. If you grew up in the US, there are things you don't think about, things like looking up a bus schedule. Not having grown up in the US, things like that are not very intuitive, so it was challenging.

Another challenge was that life was pretty expensive in the US. I remember the first time I bought jeans; they were five dollars, which to me was a lot of money. I had a hard time buying clothes because the first thing I would do is convert the US dollar to Kenyan shillings. It often seemed like an unreasonable amount of money to pay for clothes.

What I missed most when I moved to the US was my little brother. He was seven years old when I left. I really don't know my little brother. He was a really good rugby player, the captain of his high school team, and I never saw him play. I also have a little sister who is a year older than my little brother. I never saw them growing up. And for other things like Thanksgiving, especially at Macalester, people go home to their families. If you are an international student, you have to figure out, "Where am I going to go? What should I do?" Christmas break was another very tough one. Growing up, that was the one holiday we all spent together as a family. When you're at Macalester, people go home for Christmas, and you have to figure out where you will stay. When school started after the break, some of my friends coming back from home would ask me, "How was Christmas?" I'd say, "Oh, Christmas was great." And then they would ask, "Did you go home?" I'd say, "No, I did not go home." So, they'd ask, "Oh, did you not want to go home and see your family?" And that was hard because most international students knew that if you did not go home, it's not that you didn't want to see your family—it was that you probably couldn't afford to go to see your family. Some international students preferred to send money home instead of going to visit over the Christmas break. I think it's a human thing. You want to be around family—you want to be around people who love you, but when you're here on your own in school, it's pretty difficult.

I like my life in the US. It's been good for me. There's not that much anxiety anymore about the future, but I now have the problems of the rich world, as we joke with my friends. I have to worry about a 401K and how much savings I have. I think those are luxuries people have when they don't have to worry about security and a place to sleep and school fees for your kids. Believe it or not, compared to Kenyan standards, I am very wealthy—I can afford to have a car. Growing up, things like retirement were not things we talked about because we were worried about eating food the following day or worried about school fees. Now if you don't have savings for six months, people look at you like you're irresponsible. Those are the kinds of problems I have right now.

I am now an Assistant Professor of Computer Science at Macalester College. For the most part I can say that I'm doing well. There's nothing that I want. There's nothing that I need that I can't have. I don't have to worry about going hungry. I'm at a point where doing things like yoga or going for vacation seem like things I can afford to do if I want. And I guess I'm at a point in my life where I don't worry about the future as much. It probably helps that I have a doctorate in computer science, and, at this point in my life, I have enough confidence that if things don't work out in the US, I can go back to Kenya. I can teach at the local university there, so there's not that much anxiety. I have a good group of friends here, and a good number of them are former Macalester students, which happens to be because we probably share the same views in politics. We hang out every now and then—we have a number of BBQs and host dinner parties.

Contributions and Impact

I'm involved with a number of nonprofits. One of them is Kenya Society for Academic Advancement. Until recently I was the chair for the nonprofit. I'm very passionate about education, partly because I know the only reason I made it to India and to the US is because I was gifted and lucky enough that education came easy to me. I think if you're smart, education is a way of helping people to help themselves. I'm involved in a number of nonprofits that give scholarships to kids from disadvantaged communities. I wish I could give back more. I'm still on a work visa, which means there are restrictions as to what I can do, so volunteering for the most part is how I give back.

I consider myself lucky in the sense that I have a good education, and, for the most part, I don't have to deal with a lot of challenges that people who don't have education deals with. Some of my friends say that I appear very cynical about the US. Some say to me, "But you live in the US. You work in the US." My reaction to that is that before I came to the US, I always looked at it as this place where things are done right. For those that believe in Christianity, without sounding blasphemous, it's like heaven. It's where politicians do the right thing and government works the way it's supposed to work. Then I come here, and I find there are homeless people in the US. I remember that it was a shock to me because this is the land of opportunity. They are not supposed to be homeless in the US. I think part of my criticism is because I think the US can do better. But having said that, I go back to Kenya very frequently, and it's not lost on me how fortunate I am to be in the US because you have due process. You're not scared of people just upsetting the wrong person in power and disappearing. I think there is a lot of privilege to being in the US. I'm in the US and have the choice of being here. I could go back to Kenya and get a pretty good job there if I want. I still think the US is a great country, but it can do better. So, if I appear cynical about a lot of things, it's because I think we can do better.

greencardvoices.org/speakers/getiria-onsongo

Raul Velasquez

From: Bogota, Colombia **Field:** Geotechnical Engineering
Current City: Minneapolis, MN

"I FEEL LIKE A SORT OF A HYBRID: HALF LATIN-MINDED, HALF AMERICAN-MINDED. WHEN I AM HOME IN MINNEAPOLIS, I SOMETIMES MISS MY HOME IN COLOMBIA, AND WHEN I AM FINALLY VISITING COLOMBIA, I MISS MY HOME IN MINNEAPOLIS."

I was born on December 26, 1979, right after Christmas. As a kid, that's not a fun time to be born because you get one present for two and your friends are not around during this time of the year. I had a comfortable and fun childhood. I was fortunate enough to have a loving and stable home where learning was always encouraged. I was also fortunate to have strong female figures in my life such as my mom, sister, grandmas, and aunts as well as kind male figures like my dad, grandpa, and uncles. My parents continuously encouraged me through their examples to be learning, reading books, dancing, and enjoying life with family and friends. In our home, there were always all sorts of books and music from all around the world available.

Growing up in the nineties in Colombia was, however, challenging. The so-called "War on Drugs" was at its peak with cartel violence happening all around us. There were bombings happening often throughout the country, and it was draining and emotionally challenging to keep up with the daily news. You never knew if you were going to come back home safe.

My adolescent and early adulthood years were intellectually stimulating with the influence of my mom, a professor of molecular biology, who later went on to complete a PhD in stem cell therapy and neuroscience, and my dad, a mechanical engineer, who was always interested in science. I was influenced by all the books I had available at home, particularly the ones related to astronomy and cosmology. During high school, my parents gave me a telescope so I could watch the moon and the rings of Saturn. I wanted to become an astronomer but soon realized that it was not a practical career to have in Colombia and thus decided to go for something more practical in the STEM fields. After graduating from high school, where I was a total nerd and loved to study and do well in homework and all subjects, I decided to study civil engineering where fundamental principles of physics and mathematics are applied to build the infrastructure required by society.

While pursuing my bachelor's degree, I met two professors who encouraged me and provided opportunities to teach and conduct research in the area of geotechnical engineering. I became very passionate about this topic. I would stay until late hours in the research lab either performing experiments by breaking samples of soil and rock or analyzing the data from experiments and computer simulations. We published a couple of papers and presented in few national and international conferences, including one at MIT in the US. It was a great feeling to be able to present research that was useful to other research groups and practitioners of our profession. After graduating with a Bachelor of Science in Civil Engineering with emphasis on geotechnical engineering, and thanks to the influence of two great mentors, I decided to continue to pursue a master's of science while conducting research in the same field and same school in Colombia. As I was pursuing graduate studies in Colombia, we started an engineering consulting firm with a couple of colleagues and friends.

Moving to the United States of America

Back in 2002, my parents thought that I needed a break from all the study, work in the research lab, and the efforts of starting a consulting firm. The intent of my parents was twofold: rest and learn English. My dad had a friend in Chaska, Minnesota, and decided to reach out to see if she would host me for a couple months. She was more than happy to welcome me into her home. My "American mom" lived with her husband, and neither of them spoke Spanish—so I had a perfect English-learning environment. At that time, she was working for a company that made material testing machines and had connections to professors at the University of Minnesota (U of M). One random day, she took me to the Civil Engineering Department for a tour of their laboratory facilities, and she introduced me to several professors. I left a couple of resumes in case there was interest for some of the professors to work with me.

After three months in Chaska, Minnesota, immersed in the Midwest culture, I returned home to Bogota, Colombia, to resume my graduate studies, teaching, and research duties, and work on our consulting firm. To my surprise, a year later I received an email from one of the professors I met offering me the opportunity to work with him for an interesting road research project. The research project was intended for a master's student and was sponsored by the Minnesota Department of Transportation (MnDOT). At the time of this offer, I had another opportunity to pursue graduate studies in Barcelona, Spain. It would have involved investigating the behavior of unsaturated soil, a challenging topic in the field of geotechnical engineering.

However, I fell in love with Minnesota, especially the U of M campus, the Mississippi river, the lakes, the parks, and the diversity I observed during my short visit. I decided to pursue my degree at the U of M. Most of my friends thought I was a bit crazy choosing a cold place like Minneapolis instead of a well-known, warm city like Barcelona. After living in Minnesota for a few years, I have to say that I enjoy the winter and love the energy of the people coming out of a rough winter.

After moving to the US, there were several challenges and hardships. At the beginning, it was difficult because I gave up a lot of comfort and my circle of support. For me it was hard not to have my mom, dad, grandma, great grandma, grandpa, uncles, and aunts nearby. I missed my family and Colombian friends very much. In Colombia, I did not have to worry about cooking meals or cleaning my clothes. However, here in the US as an independent person, I had to take care of myself from planning for meals to cleaning my home and clothes. Initially, I experienced a culture shock that included the type of food available. The language barrier was another issue. I did have problems for the first couple of years communicating with people . . . and I still do sometimes. I remember having a hard time communicating and doing paperwork over the phone. It was terrifying for me to attempt to explain something over the phone. It was easier for me to explain or talk in English when I had the person in front of me. Another difficult feeling I continue to have as an immigrant in the US is that I feel like I do not belong either in Colombia or here. I feel like a sort of a hybrid: half Latin-minded, half American-minded. When I am home in Minneapolis, I sometimes miss my home in Colombia, and when I am finally visiting Colombia, I miss my home in Minneapolis.

STEM Career

My original plan was to come to the US to pursue a master's degree only and then to move back to Colombia to continue building a consulting career based on the company my colleagues and I had founded. Our situation was exciting and full of opportunities—we had our first couple of engineering projects and our own office in Bogota. However, the chemistry with my adviser at the U of M was great, and our academic production was high. We were publishing technical manuscripts in relevant engineering journals and presenting at national and international conferences. He invited me to stay and continue to pursue a PhD. Also back then, I started dating a very smart, fun, and beautiful woman who would later become my wife. My life was balanced: I was having a lot of fun going out dancing and traveling with my girlfriend, and my brain was happy to be working on challenging road research with my professor and other colleagues at the U of M. Furthermore, I was getting to know other students and building a strong network of friends from all over the world who were teaching me their life experiences, their cultures, and providing me with other perspectives from a personal and professional point of view. It was a very enriching experience that I could not pass on; thus, I decided to stay to complete the PhD and eventually settle down.

After graduating from the PhD program, I moved to the University of

Wisconsin-Madison to work as a postdoc in a top research group dealing with asphalt materials for pavement design and construction—another enriching experience as I worked for three and a half years with another strong mentor and with an extremely smart and diverse group of undergraduate and graduate students. This experience was intellectually productive with several published manuscripts in technical journals and conferences as well as technical reports for state and federal transportation agencies such as the Federal Highway Administration (FHWA), the Wisconsin Department of Transportation (WisDOT) and MnDOT. At UW-Madison, research and development was also conducted for companies such as Honeywell and DuPont, working on environmentally friendly and robust materials that can be used as cost-efficient solutions for our aging road infrastructure. Towards the end of my tenure at UW-Madison, however, the relationship with my girlfriend was not in great shape, and I wanted to return to Minneapolis where she was living to reconnect and work on our relationship as she was and still is a priority in my life. After moving back to Minneapolis and improving the relationship with my (now) wife, I started a consulting job at Barr Engineering as a geotechnical engineer thanks to the efforts of a great American friend who funny enough also introduced me to my wife. At Barr, I lead the computational geomechanics group and dealt with the design and analysis of geotechnical structures such as foundations, retaining walls, roads, and other structures. I also work on forensic analysis and repairs of landslides and on developing cost-effective solutions to the infrastructure and energy needs of our society.

I have always felt passion about science and STEM in general. My main influences have been my parents, who are STEM professionals, and my sister, who is an economist. The home environment my parents created was intellectually stimulating. I benefited from having so many STEM related books around me while I was growing up. My parents encouraged my intellectual exploration without restriction. Later on, the influence and encouragement of two professors in Colombia while I was pursuing my civil engineering career helped me continue to work with passion in STEM. The interest in academia, research, and STEM in general only grew stronger after coming to the US with the opportunity to work in first-class—and sometimes one-of-a-kind—research facilities at the U of M and later at UW-Madison. I feel quite lucky to have met and collaborated with my professors and colleagues at the U of M and UW-Madison. All of them inspired me to continue working and have fun learning in STEM. Currently, my passion for STEM is strong as it has been with the complex practical projects that I deal with on a daily basis with my very smart colleagues at Barr.

Diversity and Beauty of the USA

A very important mind-opening experience as an immigrant in the US is the exposure to diversity. Back in Colombia, I was exposed to a pretty homogenous society. However, from day one in the US, I was exposed to different ways of thinking and seeing the world. My friends and colleagues in the US come from all over the world, including cultures that I knew very little about before coming here. Lessons learned from all kinds of culturally diverse activities that I have attended or experienced throughout the years thanks to my thoughtful wife have opened my eyes about the importance of diversity and inclusion in both my personal and professional life.

What I love to do for fun includes playing soccer and dancing to all kinds of music. I have a passion for tango as it is like therapy for my wife and me. My wife has also planted in me the love of hiking and the outdoors. We do like to go for runs or extended hikes in nature—for example, we have hiked down the Grand Canyon three times already . . . there are probably many more times to come. Another passion I have is riding horses. I was fortunate to have my parents sponsor this activity for me since I was a little kid. I love the smell of horses and how strong and smart they are. Nowadays, I am lucky to find an affordable laidback barn nearby where I can continue to ride and jump with beautiful retired racehorses.

Contributions and Gratitude

As a professional engineer, I love to give back by volunteering with educational outreach activities at the high school and college levels, and donating money to cool STEM causes. I do also like to help our profession grow stronger by helping in technical committees and peer reviewing papers for different journals.

I am grateful to my wife, mom, dad, sister, other family members, and all my immigrant and American friends. Without their support I would not have accomplished my professional and personal goals and would not have succeeded as an immigrant in STEM in the US. From moral, emotional, and financial support to help with bureaucratic processes, you have always been there for me. I will always be indebted as you have helped me become a better human being and a better STEM professional.

greencardvoices.org/speakers/raul-velasquez

EUROPE

Slavonski Brod, Croatia

Dalma Martinović-Weigelt

From: Slavonski Brod, Croatia   **Field:** Environmental Science
Current City: St. Paul, MN

> "IN THE STEM AREA IT CAN BE DIFFICULT TO BE SELF-CONFIDENT BECAUSE YOU'RE ALWAYS LOOKING UP TO PEOPLE WHO KNOW SO MUCH. IT'S IMPORTANT TO BUILD RELATIONSHIPS ALONG THE WAY . . . BECAUSE, ULTIMATELY, WE FUNCTION KIND OF AS A COLLECTIVE."

I had a really nice childhood. My life started in Dubrovnik, but I remember very little of the well-known city because when I was three, we moved north towards what I consider my hometown, Slavonski Brod in the continental area. Dubrovnik is very warm, beautiful, and exciting. Slavonski Brod, the town I moved to, is really not that exciting of a place, but it's still a decent place to grow up. I grew up in a middle-class family that took nice little vacations like everyone else in summers on the Croatian coast. My life didn't, in my opinion, look too much different than any American contemporary of mine—I think our lives were maybe pretty similar.

I was in college in the early 1990s when there was a war as the result of the breakup of Yugoslavia. I actually was very lucky to be in the city of Zagreb, which was not as impacted. My family, however, was still in Slavonski Brod, which is on the border with Bosnia. There were lots of impacts on my hometown. There was a lot of uncertainty and not good times. It wasn't your usual college experience. But I was very young, and I think when you're that young, you feel invulnerable. Somehow I cruised through those years more easily than I think I would have if I was older. But I realized that I wanted to move away from all of that because of bad memories. The year I graduated, 1995, was about the time the war ended. In terms of opportunities, there was pretty much next to nothing. My family was very impacted by the war: savings, reserves, and aspects of the middle-class life that we lead prior to war basically disappeared.

Moving to the United States of America

I decided to go to the United States because I was familiar with the culture. I had been an exchange student in the US in 1988–89, and I felt that would be a comfortable place for me to go.

My long-term plan was to win a scholarship, move to the United States, and go to graduate school. I received a scholarship from Central European University, and then the government shutdown happened, which stuck me in Croatia longer than I thought. There were all these last-minute expenses, which left me to travel to the United States with a very limited amount of cash. My flight got stuck in Amsterdam, and my cash pile was dropping down. When I finally arrived in Memphis, I was separated from my suitcase, which in the 1990s, was a very different thing than today. The chances of getting together with your suitcase were very small, and I hated that because I packed only one suitcase. It was packed with the things that don't make sense—not necessarily things you need but the things that you must have with you. I had momentos and photographs.

When I arrived in Memphis, I was very lucky. I had a boyfriend who was also of Croatian descent, and he had just started living there. We went to this bar near the college—there's no better way than to start your life in the United States as a college student than to have a beer with others. That's where I spent the last of my cash. I did not realize that my cash is not going to be in a bank account because of the shutdown. For a few weeks I lived without my suitcase and a minimal amount of cash. The suitcase arrived eventually, but by then I had adjusted to this new life without my things and money. I had to be patient because I knew there was that chunk of money that was going to help me during my first year here. I know that's not the case for everyone, so I felt that my arrival was very easy in comparison to many others I know.

I moved to the US alone. I had a little bit of support when I moved, but I gave up close contact with my family. I missed out on many years of friendship and family time, and I rarely speak my language from home. But maybe the thing that I gave up the most is a sense of clear identity. It's a very strange place to be in when you go back to Croatia and people viewed you as the "other" . . . and then you're also the "other" here. So you're kind of in this in-between space, which is a very bizarre but liberating space. So you give something up, and you gain something.

I arrived a little late for classes, so I couldn't start courses on time like the other students. I had to make up for it, but I was always a little off on all of the timelines. In that first month, my neighbor Bill taught me how you function in the United States. I remember learning how to write a check. He helped me figure out how to get the phone and water installed in the house—those kinds of practicalities of setting up life. Also I remember not having any furniture, which was bizarre. I would have probably never had that experience in Croatia—people were dragging all this old furniture over and bringing it in. The place was just a medley of very strange items. The first month was mostly filled with excitement and learning a new life.

Collaboration, Key to Success

One of the things that was really important for my success was collaborations with people I met along the way. I really attribute my successful career to collaborations with individuals, and I found Minnesota to be a great place to collaborate, especially in STEM. It's a very rich place with academics and universities. It's also a very rich place with nonprofit organizations and state and federal agencies. Minnesota is a unique place to work, in my opinion. I've been able to benefit from the interactions and collaborations with other organizations, and I have been able to give back in that same way. For example, I work with undergraduate students that like to do science. I had a few students work with some of the nonprofits in the Twin Cities.

One nonprofit is called Youth Farms and works in youth development, and we worked with them to incorporate a science-related project. What I love the most and what I learned about STEM in particular is that I think the way to be successful is to surround yourself with a diverse group of people who have a variety of experiences and expertise. Often people think to affiliate with people who are really famous, but it turns out that the students that I work with have been a really rich source of new relationships and projects for me. I have students who went back to Croatia multiple times who I find are inspiring to reconnect with. Some have projects in Croatia right now. I have also been really motivated by one of my colleagues who was at the University of St. Thomas; I have also had students going there.

STEM Education and Career

I grew up in a family that likes humanities, and my dream was to be a comparative literature scholar. I never really liked science: I just happened to be good at it. For very practical reasons, I was actually going to go to medical school and study engineering. What transformed my experience, oddly enough, and made me want to be a scientist was two things: I really love nature, and I pursued biology because that's something I have a connection to. When I was in the United States as an exchange student, I took chemistry, biology, and physics courses with high school professors who really left an impression on me. They basically showed me about the process of science. That's something I wasn't familiar with until then. It's a whole different story to actually get engaged in the discovery process than to learn and memorize facts. My high school professors did such a beautiful job of it, and

I was like, "Yeah, this is actually interesting, and I could see myself investing in that."

One of the challenges if you're an immigrant is that you have to have the next visa for when you are going to go into your next job, which means that you have to have your next job lined up very quickly before your prior job has ended. You don't have many choices compared to someone who has US citizenship. I started at the University of Mississippi, and I finished my master's. Then I needed a job very quickly, so I applied nationally to all kinds of jobs, but I failed to get jobs because I didn't have lots of work experience or the help of a network. I think because you are an immigrant, people don't really understand what your background is or your challenges. Many organizations couldn't provide the support to hire someone with my high level of education.

I ended up having a really cool job after my master's. I got the job at the medicinal plant garden at the University of Mississippi, which was a part of the pharmacy school and Center for Natural Products Research. Most people would know that place as the only place at the time where marijuana was grown for research reasons. It was a bizarre first work experience, but I learned so much. It was interesting because it wasn't my immediate field of expertise, but because that was the first job that opened, I took it and learned many skills that I find myself using now. I learned what the process of natural product discovery looks like and what the flow of the science is. I also made some really valuable connections during that time.

Through a series of windy paths and experiences I moved to the University of Minnesota. I proceeded after my PhD to go to work as a postdoctoral researcher at the National Academies of Science, Engineering, and Medicine on a project with the EPA, Environmental Protection Agency; however, because I'm not a citizen, I couldn't actually get a job there after my postdoc. I had to leave and look for jobs outside the government, even though that would have been my personal first choice. That's how I ended up at the University of St. Thomas. They offered me a position, and there was a six-month delay in my ability to get all the paperwork in place to start there, but they were so generous and patient with me. I know that not all organizations are like that.

I live in St. Paul. I like the pace of it. I don't drive, so I walk around a lot. Most of my free time revolves around spending time in nature, outdoors. I also read a lot and enjoy time with friends and my husband's family. I am married to a Minnesotan who also enjoys the outdoors very much. It's been a really pleasure to acquire this other family here. My work life as an environmental scientist often percolates into my private life. I always want to go somewhere where there's something I'm curious about. We lead a boring, calm life that we enjoy very much.

Contributions and Impact

My main contribution to Minnesota has been through discovering new phenomena in methodologies that will help protect the environment. I work on several projects, and as an environmental scientist, I'm very interested in water protection. A legislative commission of the citizens of Minnesota selects projects that target issues that are important to Minnesotans and natural resources. One of these projects relates to unregulated contaminants in the environment, and the other one relates to how old oil spills are aging. We're trying to advance our understanding about whether water is safe now and in the future. We have funds to conduct research that will promote healthy drinking and surface water. Most of my work is in understanding the water quality and how we can protect the water for future generations, not only in Minnesota but also in other places.

In the STEM area it can be difficult to be self-confident because you're always looking up to people who know so much. It's important to build relationships along the way with people who know more than you do and people who will learn from you because, ultimately, we function as a kind of collective. When it comes to science, it's rarely a lone scientist making amazing discoveries. There's so much richness and innovation that can come out of collaboration. I think the most important lessons I have learned are to be confident, be curious, and collaborate.

greencardvoices.org/speakers/dalma-martinovic-weigelt

AFRICA

Mogadishu, Somalia

Hussein Farah

From: Mogadishu, Somalia

Current City: St. Paul, MN

Field: Software Development

> "THE MORE I CAME TO LEARN ABOUT TECHNOLOGY, THE MORE I REALIZED THIS IS A GOOD PLACE TO BE BECAUSE YOU CAN IMAGINE NOT ONLY WHAT MAKES YOU ANXIOUS BUT ALSO IMAGINE WHAT CAN MAKE YOUR LIFE BETTER. THERE ARE ENDLESS POSSIBILITIES."

I was born in Somalia. I had a fairly normal childhood up until the civil war disrupted our lives. Like most Somali refugees, my family and I were violently uprooted from our home country to Kenya. It was a very abrupt process. In the middle of the night you're just being told, "Get out! Get out!" And you have to get out. We were very anxious as a family—my parents were extremely nervous. My first couple of years of formal education started in a refugee camp in Kenya. I grew up in Kenya, where I did my middle school and high school.

Moving to the United States of America

I was very fortunate to have some scholarships that took me outside of Kenya for further studies. I came to the US to do my master's in systems engineering. When I was moving to the United States, I was a young man. It was my choice to come to the United States because I came to do my master's here through an asylum/refugee program. At that time I was mature enough to be actually looking forward to my arrival to the US.

When I was coming to the United States, things were moving so fast. When you are being told there is a scholarship for you and a process for you to go to the United States, you only have "X" amount of time to react. You only have a short time period to respond. The only thing I remember is that I was very excited to make the move. I was really looking forward to it because everybody thinks about the United States as the place to be. So I was like, "Yeah, okay, that's where I'll go, and I will look forward to that."

There was nobody that was waiting for me in the US—I came alone. I remember that I had only fifty dollars in my pocket. I did not know anybody, and it was overwhelming to some extent. It's a weird place to be because you don't know what tomorrow will bring, but yet you are looking forward to tomorrow. I remember vividly, just thinking, "This is a big city, with a lot of people, and yet I don't know a single person. What will tomorrow bring for me?" And I just kept telling myself, "We'll see. We'll see what comes tomorrow."

STEM Education and Career

Initially, I did my undergraduate studies in international finance. As the world was transitioning from 1999 to 2000, the whole world was panicking with the Y2K phenomenon. At that moment, information technology (IT) captured my imagination. I thought, "Why is it that everybody thinks that their life will be disrupted? And what is it about IT that there is a possibility that your life will be disrupted?" I saw myself gravitating toward IT around that time. I did a lot of coursework in software and systems engineering, and the more I came to learn about information technology, the more I realized this is a good place to be because you can not only imagine what makes you anxious but also imagine what can make your life better. There are endless possibilities. If you think about what we have now, we did not have this technology five years ago . . . and what we had five years ago, we didn't have ten years ago. And maybe what we don't have now, we will have five years from now. It's all about IT. Everything is in flux . . . everything moves now . . . that's what excites me about technology.

After a few years in Washington D.C., I read an article that really made me think about my purpose in life. It was an article in The Washington Post about the plight of African immigrants in Minnesota. It was talking about how there is a large Somali population in Minnesota, but they were having a very hard time adjusting to life here. In particular, many of the new communities were experiencing difficulty with the financial culture in the United States. I felt like having reached that level of education myself, it was up to me to give back, to go back and see how I can make the lives of other African immigrants better.

That was why I moved to Minnesota. There was a very distinct connection that I got from reading that article that made me go to my boss in Washington D.C. and give him a two week notice. He never understood why I did that. He said, "Why are you going to disrupt your life in D.C. and move to a place that you don't know anybody? You have nothing lined up and yet you are willing to go there." And to his point, I didn't know how cold Minnesota was. Now, twice a year, I second-guess myself. I wonder why I'm here, knowing that we have negative fifty-five degree temperatures. But that opportunity, that distinct connection, only comes to you when you are an immigrant yourself, and nobody can teach you that. Nobody can make you do something unless it's in you. That is a connection, a bond that I got from just reading this one article about Somalis in

Minnesota. I drove for twenty-two hours nonstop from Silver Spring, Maryland, to Saint Paul, Minnesota, and started my life here.

Uplifting A Community

I did a lot of work in IT as a systems analyst and as a software developer, and to some extent I was enjoying that, but as a community, I've felt like communities of color were not participating in this space. If you think about the IT sector, we are all in the 21st century digital economy. The whole world is moving towards that. But if you think about who is there and who is not there, ninety-nine percent of us as immigrant first-generation people of color in Minnesota are not in the high-tech space, and everyone else is moving towards that. It almost became a personal calling for me. If we really want to uplift our communities, I should not be comfortable just making the money myself—I should be thinking about more of us moving into this sector. That's why I founded the New Vision Foundation in 2016. Our mission is to create pathways to success for disadvantaged youth in Minnesota ages thirteen to eighteen years old through coding and digital literacy classes. We are now teaching over one thousand students in about twelve schools across the Twin Cities metro area. New Vision Foundation is educating the future STEM workforce in Minnesota.

If we, as a society in Minnesota, don't make a very deliberate effort to make sure we all participate in STEM, a big chunk of our community is going to be left behind. We are in the 21st century digital economy. Everything is moving towards tech. If we think about five years ago, if you went into any major retail chain like Walmart or Target, you were most likely going to have a cashier, but now you're most likely going to be on a self-pay, automated checkout lane. If you think about companies like Uber, Airbnb, and Lyft, they were nonexistent five years ago, but now everything rotates around them. If you think about Minnesota communities like Marshall, Wilmar, or St. Cloud, the reason why there were large immigrant populations moving to these cities was because of factory jobs. Now, those factories, in their own self-interests, are automating their processes. In an assembly line you used to have twenty people. Now, there's only five people working at a time. Fifteen jobs have been eliminated because those jobs are now automated, and you cannot fault the companies for wanting to improve their bottom line. But you ask yourself, "Where are those fifteen people going to get jobs again in those cities?" They'll be forced to move out.

If more immigrant populations move out of those small cities like Marshall, Wilmar, and Mankato, those cities are going to shrink. The macro systems will punish us for not making sure that the communities are all engaged.

Then you look at the racial income disparities—we are not participating in those higher paying jobs. Right now, Minnesota is ranked one of the worst states in the US when it comes to racial income disparities. People that look like me don't make as much money as their mainstream Caucasian peers. If you really want to close that gap, we've got to have more people of color in those high-paying jobs. If not, the gap will continue to increase, and that is not sustainable. We cannot have a state that has people who *have* and people who *don't have*.

Contributions and Impact

Now I live in the East Metro. I have a family—a wife and three kids—and I love playing soccer and listening to music. My contribution to my community is just to actively work on making sure that Minnesota is a society for everybody and to be proactive in engaging, motivating, and inspiring communities of color to enter the high-tech sector. I am not thinking about the individual trees—I'm thinking about the forest. How can we help more of us participate? Because if we have more of us participating, then in about five years from now, you can think about most of us having high-paying jobs and contributing to the welfare of the whole state. Then, the health of the state becomes much better. This is my personal calling. I look forward to every single day, going out there and educating young kids on the importance of the high-tech sector through coding and digital literacy classes. I'm making them get excited and making them see themselves as creators of technology. I want people to shift their mindset from being just users of technology to creators of technology. Once you make that shift, then everything is possible.

greencardvoices.org/speakers/hussein-farah

ASIA

Phnom Penh, Cambodia

Kim Uy

From: Phnom Penh, Cambodia   **Field:** Public Health & Surgery
Current City: St. Paul, MN

> "I CAN COME TO THE UNITED STATES. I CAN BE WHATEVER I WANT BECAUSE THAT'S ALL YOU PICTURE ABOUT AMERICA. IT IS THE LAND OF THE DREAM AND YOU GO THERE AND THEN YOU KNOW YOU CAN ACHIEVE WHATEVER YOU WANT AND THAT'S SORT OF MY REACTION TO COMING TO THE UNITED STATES."

I grew up in Phnom Penh, the capital city of Cambodia. I feel like the way that I used to remember Cambodia has changed so much. I remember my house and my three brothers and going to school with my friends. I remember splitting time between morning and afternoon going to two different schools. In Cambodia, if you're of Chinese descent, it's culturally required for you to go to a Chinese school. My mom is of Chinese descent, so I was required to go to a Chinese school, in addition to Khmer school.

In the morning, I remember getting up early to go to Khmer school that started at 7:00 a.m. I would walk to school each morning, which was about ten to twenty blocks from my house. Khmer school is in the public system, and it's not very good. I remember showing up to classes, and the teachers were there, but they didn't really teach us. I would spend a lot of our time just sitting in a classroom chatting with all my friends, so that was kind of fun from my young self's perspective. Then when I was done with school in the morning, I would come back home to shower and get clean before having lunch and going to Chinese school in the afternoon. It is hot in Cambodia, so I had to shower often to get clean. Chinese school is very highly disciplined and more strict than Khmer school. The teachers didn't really like us chit-chatting too much in class or talking loudly. The teachers would teach us many subjects in Chinese language, so that was a difference in the public and the private systems in Cambodia.

Comparing that experience to my high school education in the US was interesting. I remember the US high school class was structured similar to the private Chinese school. It was strict, and no talking was allowed during classes. The materials that were being taught in the US were less advanced compared to what I had learned in Chinese school in Cambodia. For example, I remember having already learned algebra and geometry in sixth or seventh grade in Cambodia, but in the US you're not supposed to learn them until you're in ninth or tenth grade. I think the public education in the US is behind the education provided in the private Chinese school.

Moving to the United States of America

My aunt, who is a US citizen, sponsored my dad and our family to come to the US beginning in 1991. It took ten years to go through the process and be on a waiting list. In 2002, my parents told me and my siblings, "We are going to the United States." The news was sort of a surprise because I didn't have a lot of decisions in that process. It then took four more years before we were able to get to the United States in 2006. I remember having to go to Thailand for the visa interview before we could come to the US. I remember a part of me was feeling sad because my youngest brother was not coming with us—his process was delayed because he was added after the original paperwork before he was born. It took another five years before our family was reunited in 2011. Another part of me was also excited. I remember thinking to myself that I can come to the US. I can be whatever I want because that's all you picture about America. It is the land of the dream, and you go there, and then you know you can achieve whatever you want, and that's sort of my reaction to coming to the US.

My family came to the US by plane—that was actually my first time being on a flight. I remember I was so excited. We went through Taiwan to get to Minnesota. It was a really long flight, probably fifteen to sixteen hours, but I stayed up the whole time because I was so excited to be on a plane. I was feeling like, "Wow, I'm in the air." I remember being seated in the middle aisle of the plane, and I couldn't see through window, so I had to get up and walk to where you can look out the window and see the sky. I remember being excited about that.

We arrived here in June, and it was warm and summer for the US. But for us it was like, "Wow, it is cold here." I remember we bought winter jackets in Cambodia knowing it was going to be very cold, and we wore them! Looking back, it was very funny. On my first day in the US, I remember feeling very tired because I hadn't slept in a long time. My picture of America was very different than what I saw in Minnesota. I imagined America would have a lot of tall buildings, a lot of very highly-populated areas, lots of people walking around the street, and people speaking English everywhere. But then when I got to my aunt's place in Shakopee, which is a suburb, there weren't a lot of tall buildings or anything like that—and there were a lot of farmlands. I remember thinking,

"Wow, they farm in America!" That's weird because I would not have imagined it. Cambodia is a country that focuses on agriculture, so we do a lot of farming there. I never realized that in America they also do farming—that was strange to me. I remember my uncle took us to the Mall of America, and I thought it was such a big shopping mall. I've never seen any shopping mall like this before. I remember there is a rollercoaster on the inside. That was my first time seeing a rollercoaster and first time seeing a mall.

I think my parents decided to come to the US, from my interpretation of their feelings, because they felt like it was a better opportunity for us. My parents owned a rice wholesale business. I don't think any of my siblings and myself would like to inherit my parent's business. My parents were really concerned about our future. The Cambodia job market was really bad, and they weren't sure what my brothers and I will be doing if we weren't inheriting their business. I think it's more for us than for them that my parents decided to come to the US.

I don't remember that I had to give up much to come here. Though reflecting back, I realize how much my parents have given up for coming here. My parents gave up their business and left all their family, friends, and their social networks behind. They also had to leave our house. I remember a couple years before we came to the United States, my parents had just bought a house that they were really excited about because they were finally able to afford a comfortable lifestyle. I remember they had to sell our house when they decided to move here. They also had to give up being business owners in Cambodia. Now they are working in an assembly line of a food package company. It is a hard job. It is different for them because they used to be a boss, but now they have to be something else, a worker. So it's hard for them.

Learning English

I think the most challenging part of being here is learning English. I didn't know English beforehand. All I knew was my ABCs, A through Z. I remember it was hard for me to learn English. While in Cambodia, I remember it was really difficult for me to learn Chinese—even though I started when I was six or seven years old. Now, at fifteen, I had to learn English. I remember not feeling confident learning or speaking English. I remember when we would go to a grocery store, my parents would sometimes ask me for help when they needed to find something or needed to talk to a sales associate because my English was slightly better than theirs. I would always feel very uncomfortable and embarrassed having to speak English in front of my parents, in front of everybody—just speaking English in public.

I then realized that my parents were probably having a harder time learning and speaking English than me. I remember seeing them going to ESL classes every morning. At the time they were working overnight shifts. Their shift would start at 11:00 p.m. and end at 6:00 or 7:00 a.m. They would come home, eat breakfast, shower, and then go to ESL classes at 9:00 a.m. I saw them doing that for a couple years when I was in high school, and I realized that if they could do that, I should be able to learn English. So learning English was an obstacle for me, but my parents helped through that obstacle. Now I speak English much better than my parents.

I was always interested in science starting probably in high school. The high school that I went to had this particular ESL biology class for people who don't speak English, which had a lot of hands-on activities. I remember one time using this spinning device—now I know it's called a centrifuge—to get DNA out of our saliva. I thought that was the coolest thing because I never knew that you could do something like that. I think this class helped spark my interest in science. Later in high school, I was really interested in forensic science, so I decided to join the Army Reserves hoping I can get some law enforcement experience and pursue a career in forensics. But then I realized that being in forensic science, your work is to catch a criminal, and I realized I don't really want to do that. I want to use science to help people and to help make things better rather than catch a criminal.

STEM Education and Career

After high school, I decided to go to the University of St. Thomas. At the time, I started taking classes and became interested in neuroscience. I remember taking one class and learning about neurotransmitters in the brain. It was very fascinating to me to learn about how there are chemical compounds in the brain that interact with one another to drive our behaviors and emotions. Because of this I decided to major in neuroscience. Then I started working at a lab at the university where I looked at how environmental factors can affect animals' brain structure. I learned about the traces of estrogen in our water environments and how this links to a change in brain structure of animals living in those waters. These findings sort of capture the idea of social determinants of health; the idea that our social environment can affect our everyday life, shape the way we are, and impact our health. I think this was the beginning of my interest in public health and medicine. I served in the Army Reserves for six years, and that experience helped me grow in a way that I didn't realize until I reflected on it. I remember when I first joined, my thought was that I wanted to go into law enforcement and being in the military would look good on my resume. Secondly, I joined the army for the tuition assistance program. I know my parents weren't able to support my college tuition, so I need to find my own way. After six years, I got more from the army than just these two goals. My experience in the army has helped me realize that although I'm just this small Asian girl, I can do a lot. I can do anything that other normal, average-sized Americans can do. I can carry my gun. I can carry all the equipment. I can do long walks with thirty to forty pounds of equipment on my shoulders. I can achieve what I put my mind to—I am self-sufficient and have self-determination.

I decided to apply to medical school during my junior year, but then I realized I'm not competitive when compared with my pre-med peers because of my grades in my science classes. I remember I really struggled academically. I couldn't read textbooks. It not only put me to sleep, but I just did not understand what was being said to me or what was being written. I was having a hard time just trying to read and study for science classes. I thought, "How I am going to apply to medical school if I can't even read a textbook in English?" That was really difficult. I got a lot of support from some of my teachers, and that was inspiring. They guided me through the process and told me, "You can do this." They told me there are others that have come before me and if they could do this, I could too! So I decided to take a year off and study for the MCAT, which is the test you take for medical school. This gave me time to study for the test, and because I had a decent score, I could actually put my application together and apply to medical school. I ended up at the University of Minnesota Medical School because that was the school that accepted me.

I think the main lesson from my journey is that a lot of people are going to look at you and tell you it's not a good idea to do XYZ, but I learned that these people aren't truly evil. Actually, some of my mentors and family members, whom I have great respect for, would tell me, "Medicine is not for you . . . don't go to a four-year university because you don't know if you're going to make it . . . it's very competitive . . . try out a two-year degree first and see if you can handle the workload, then you can go to a four-year college." When I first heard those words, it completely crushed me because I felt like they didn't believe in me and didn't believe that I can do all these things. Now looking back, I don't feel like they were terrible people when telling me those things. I don't think they wanted to crush my dream. I think they told me those things to warn me. I feel like a lot of times people would tell you this thing because they feel like they went through something similar or they know someone who went through something similar and the idea of not being able to pursue your dream is terrifying to them. I think that's why they tell you that it's not a good idea to pursue your dream. In a way it helped fuel my energy. I had never been mad at them, but just wanted to prove them wrong. I think if you believe in something, then others will believe in it too.

Contributions and Impact

Recently I took a trip to Cambodia to do a two-month surgery rotation because I'm very interested in becoming a general surgeon with a particular interest in practicing in low-resource settings. Cambodia is a low-income country, and they have a lot of issues with surgical sourcing and workforce capacity. They require a lot of help from the international community to come in and help with their health care system. It was very interesting for me as a Khmer American to go there and learn about the Cambodian health care system. I was very interested to learn how the international community and Khmer doctors interact with each other and how they provide care to the Cambodians, especially the Cambodian poor, who are the most vulnerable because they can't afford any medical care and have to travel far distances to get any kind of care or surgery. To learn all that and contribute to the care of Cambodian people was very rewarding for me. My goal for the future is to do more—not just going back and doing annual mission trip type of visits but rather creating something more sustainable, something that can help the Cambodians to become more self-sufficient rather than relying on the health care workforce of the international community.

Today, I am married to my husband. We've been married for over a year now. Professionally, I'm a student in a dual degree program at the University of Minnesota. I will complete my Medical Doctor (MD) and Masters in Public Health (MPH) in 2020. After that, I will go on to my general surgery residency. My particular interests for my professional career are to work in surgery and public health research, particularly in environmental sustainability and cost-effectiveness research. Currently, I'm working on a project that aims to reduce waste in the operating room. Working on this project gave me insights into hospital waste management. It's interesting to learn that hospital is one of the major generators of waste in the United States. The operating room is contributing probably thirty percent of that waste. In the era of our throw-away society, we sometimes think it's okay to have all this disposable waste. But to me, I think it's important to think about how we can minimize our waste and think about the impact of our behavior in our society and on climate change. I feel like this is something we can do better—we can improve the process of care and overall efficiency by minimizing the waste. Health care is not only about helping the patient who is being treated, who is on the operating table, but also helping everybody globally.

greencardvoices.org/speakers/kim-uy

AFRICA

Eket, Nigeria

Itoro Emmanuel

From: Eket, Nigeria
Current City: Sartell, MN

Field: Nursing

> "I LOVED BASKETBALL . . . WE HAD A PUBLIC COURT THAT WAS ABOUT FIFTEEN OR TWENTY MILES AWAY . . . AND I WOULD WALK FIFTEEN MILES TO GO PLAY BASKETBALL . . . I WOULD DO THIS DURING THE HOLIDAYS WHEN I DID NOT HAVE SCHOOL, FOUR TO FIVE TIMES A WEEK. MY DAD THOUGHT I WAS CRAZY, BUT IT WAS SOMETHING THAT GAVE ME A LOT OF JOY."

I am the first son of three boys, the middle child of six children. I have three older sisters and two younger brothers. I grew up in the southern part of Nigeria in a town called Eket in the Akwa Ibom state. I lived there until I left for college. My mom and her sister had decided that they would raise each other's children because they did not think they would be disciplined enough with their own children. I lived with my parents, but in elementary school and middle school I lived with my aunt. I went to elementary school on the Nigerian navy base in my hometown, which had a school for servicemen and women's children with a few slots made available to members of the community.

Attending that school meant I had to move to a different state to live with my mom's sister and her family, and we'd come home at the end of the school year. When it was time to go to high school, I came back home because though I was originally supposed to go to a military academy in the northern part of Nigeria—my mom couldn't handle me going to school that far away. The distance would be equivalent to living in Minnesota and going to school in Georgia. I ended up going to a Lutheran high school in my home state. After high school, I attended the Akwa Ibom State University majoring in electrical engineering for the first three years before transferring to the US to finish college.

The small town I grew up in was a coastal area with a lot of diversity and good job opportunities. Mobil petroleum company had a drilling rig with an onshore base, so all the people that worked there came through the town. There were a lot of different tribes that lived in my town, and we were exposed to some of the white men who managed the companies. Some of them might have been British or American. Tourists were seen and treated as precious commodities— you didn't really get close enough to touch them, but you could see them from

57

a distance. They lived in a fenced-in compound with very strict access to the general population. We would see them riding the company bus to work. Because we lived in a coastal area, fishing was a mainstay for young men and women who were not lucky enough to get a job working in the oil companies. Working for the oil company was the best paying job in the state, which meant most people who went to college majored in engineering. Thinking about it now, I realize that depleted the talent pool for everything else because everybody wanted to be able to make a good and decent living.

I loved basketball. We had a public court that was about fifteen or twenty miles away, so I would get up, pack my jersey and shoes in my bag, and I would walk fifteen miles to go play basketball. My home was the farthest away from the court, so along the way I'd pick up my friends. We would come back home the same way, and the group would just simmer out until there was just me walking the last five or ten miles by myself. I would do this during the holidays when I did not have school, four to five times a week. My dad thought I was crazy, but it was something that gave me a lot of joy.

Moving to the United States of America

My eldest sister moved away first. She moved to London. Her fiancé at the time had convinced her to move. I was always fascinated with the US. There was something about America that was intriguing to me, so when a friend of mine from Nigeria brought up the idea of going to the US for college, I thought it would be a cool idea to go to the US and study and then come back. I thought maybe I could then be a boss in an oil company. That was my plan; that was always my plan.

There were four of us—me, my friend, and his brother and sister—applying to schools in the US together. When I got admitted, I told my parents, and they were supportive. My friends who also applied and I had to drive to an interview at the embassy, which was in a different state. Each person had a different interviewer, and we met up at the end after we had all been interviewed. I was the only one of the four of us who was given a visa to come study in the US.

When I told my dad about the visa, he was very excited. My dad was someone that believes so much in education. He didn't have a formal education—I am not sure if he finished high school. Because my dad believed in education, if you wanted anything from him, you had to tie it to school. If it was related to schooling, you had a ninety percent chance of getting what you were asking for. I knew telling him that I wanted to go to the United States wasn't going to be a problem, but it was the financial obligation that made me nervous about telling him about it. It was going to be expensive when you look at the tuition and everything you needed. My dad was more financially stable than most other parents around where I grew up. He started a construction company or more like a gravel company. Shortly after he started the company, Nigeria went through its independence, and they were rebuilding roads. He made a lot of money from that. After that he got into the electrification business; he got a big contract from the government to electrify the states, set up transformers, and run the wiring. He made some good money then too. By the time we all grew up, he wasn't really doing much of that because he had retired. So, I knew he could come up with the money, but it was going to take moving a lot of things around to make it work.

When I told him I was going to the US to study, he was very proud and happy. I explained to him that I still needed a visa and admission did not guarantee that I would get one. He felt it was a waste of time then. He said, "If you're not going to get this, if it's not guaranteed, why are you wasting your time? You are in college now; just focus. Get your degree. Don't chase something you might not be able to achieve, I'm not putting my stock on it." In my community there are not a lot of internationally educated people, so I had like a thirty-minute window of him being excited about it. However, when I went for the interview and got the visa, the told me, "Let's do whatever we have to do to get you up and going." He ended up selling one of his houses to a friend of his so that I had enough money for my tuition, housing, and everything I would need. I left right after that, but I never got a chance to thank him because he died about a month and a half after I got to the United States. I didn't know how to use the phone to call Nigeria yet so I couldn't talk to him or my family. At first we just emailed. I was very close to him, but I did not hear about it at first because my family didn't know how I was going to react. I learned about it later from a friend and then talked to my brother. I still haven't really been able to mourn him. When I traveled back to Nigeria for the first time, I went and saw his grave but that did not give me closure. I can't blame them for it; they were trying to do what they thought was best for me. I honestly don't know if I wanted to know right away. I think the way I found out about it eased the blow, but then, I have no closure with that matter.

I had two months to get ready for travel to America and say goodbye to friends and family. I was going to Indiana, but I had never heard of Indiana. When you think about the US, you think of California, Texas, or New York. I arrived in Chicago on January 6, 2004. While waiting at O'Hare airport, I wanted to go and see what it looked like outside. I was only wearing a shirt and jeans— no jacket, no sweater, nothing. When I left Nigeria it was about eighty degrees outside. When I walked outside and opened the door in Chicago, it was blowing snow and freezing! Initially, I was just mesmerized by the snow—the cold hadn't hit me yet.

I didn't know anybody in the US. I had an uncle that lived in Connecticut, but I couldn't get a hold of him before I left. I got off the plane at the airport in Indianapolis, walked outside thinking I would grab a taxi and negotiate the fare like you do in Nigeria. I called a taxi and told him that I needed to go to Indiana State University in Terre Haute, Indiana. He drove me for what felt like forever, and when I asked how much I owed him, he told me, 140 dollars! I did not really grasp the concept of how much that was, but I knew that was all the money I had in cash left on me. I didn't know how bad of a bargain that was until I spoke to somebody a couple of days later when I learned a bus or shuttle was what I should have taken. That was my first experience: "Hey! Welcome to America!"

When I first got to Indiana State University I was told the great Larry Bird, the Hall of Fame basketball player from the Celtics, was an alumnus of our university. I was like, "Hell yeah! Basketball!" I loved basketball, and the school was big on basketball too. Perfect match. There were some freshman international students at my dorm that were also into basketball, which created a bond and friendship for me. We would leave campus sometimes to go play pickup basketball in the community. There was a part of town, the west part of Terre Haute, Indiana, and one of the first things anybody told me was, "You can't go there. . . . It's a suburb of the town . . . It's mostly white working class that live over there. . . .You can't cross that." There's a little bridge area that leads into the suburbs. . . . It's only for white folks." And I finally said, "Okay," still not comprehending what the big deal was.

I had a job at a little factory right next to campus, and all I did was take two old CDs and put a sticker on it that said, "two for the price of one." That was the job that most college kids would do. It paid minimum wage, but I was excited for a little extra money on the side. But any time we would go in—there were about four black kids that worked there, and we would stand right next to each other, as a comfort thing while working. The supervisor would come and separate us. At first it didn't really make sense. I'm like, "We're not really talking. We're not really doing anything. Why won't they just let us stand next to each other and work?" We were talking to one of the African American students on campus about the whole working situation, and he said, "Yeah, we don't go to that factory. We don't work over there because they treat you like animals. They think you're conniving, and you're trying to steal something." I come from Nigeria. We're all pretty much black. There's a little bit of tribal conflict, but I didn't really understand the concept of what racism really was or how I would get treated different just because of the color of my skin. I went to visit my uncle in Connecticut during the summer break where I got a job at a milk crate factory. The working environment was different there—African Americans, Africans, and Hispanics all worked together. I did not notice what I witnessed in Terre Haute. Some other towns are not as bad—it's just where you are.

When I came back to Terra Haute for my fall semester, a kid from Africa told me he was applying to St. Cloud State University because they have a scholarship for international students. I found out that if you are accepted, you only have to pay half the tuition. At this time, I was beginning to sweat about my tuition. My mom was supporting three daughters and three sons. So I was like, "Alright, anything that can help financially, I'll go for it!" I applied to St. Cloud State University, and I got it. That's how I ended up at St. Cloud beginning in Spring 2005, and I never left. I also decided to switch my major after I found out more about what happened to my dad.

STEM Education and Career

I knew my dad had diabetes, but he never really took care of himself. My dad wouldn't go to a doctor unless he was dying. He'd been battling diabetes with just diet. He tried some medication for a while but didn't stick to it, never checked his blood sugar, didn't really understand the concept of managing his diabetes—just because you are not running a fever doesn't mean you were not slowly killing yourself. I knew three Nigerians in nursing school at Indiana State University, and they would talk to me about things they were learning in school. They'd come home and ramble about nursing concepts, you know, all the new exciting things they were learning. So when I moved to St. Cloud State, I knew I wanted to go into nursing. I figured everyone in my family goes into engineering, but I could better help them if I could at least talk to them about health, especially because the life expectancy in Nigeria is pretty low. I think it's like sixty-something. I recalled about fifty percent of the kids I graduated high school with, had buried a parent before they were done with college or right around that time. It was either a heart attack, diabetes, stroke, or something traumatic. It didn't really register to me until I got here and my dad died. I also realized I can get an LPN degree and be able to work at a better job.

In St. Cloud I remember living in a four-bedroom apartment with four complete strangers. We all rented out a room in the apartment. We all shared one kitchen and shared one bathroom. I was never there unless to sleep. When I met my wife, Kelly, and we got serious, we moved in with her mom. I spent most of my days at her mom's house. When I was almost done with my LPN degree, we got an apartment together. I had saved up some money, and I was able to work more now that school was over. We found an apartment and moved in, and she got pregnant. By this time I haven't told my mom how serious my relationship with Kelly had gotten. When she got pregnant, I was like, "Oh no, I have to tell my mom." I wasn't going to tell my mom I got somebody pregnant out of wedlock, so we got married. We had to get married if we were going to have a baby—African tradition. We have to be a family so I can tell my mom that we're having a baby. Kelly changed my life. She made it better in more ways than one.

I decided to transfer to a technical college, Itasca Community College, which was about three and a half hours away from home. It was a hybrid program, mostly online, but I had to go up there to do my clinicals. I could do my schoolwork while I was at work, which was mostly over nights. Kelly was also going to school at the time for political science at the technical college, but we couldn't both be in school and work, so we agreed that when I finished, she would go back to school. Our first daughter was Taisha; Kelly had Taisha before we met. Then, we had Eliora. We bought our first house in Sartell. I then went back to school and got my RN degree. I came back to St. Cloud State for my bachelor's degree. Kelly went back to cosmetology school. I'm taking a break right now for maybe a year and a half, before I start back for my doctorate. I've been going to school ever since I got to the US.

Contributions and Impact

I'm a nurse, and I went into nursing because I care about people. I love helping people. I get great joy when I can care for somebody and give them an amazing experience. They might never remember me, but they will always remember how I made them feel. And that's one thing that gives me the most satisfaction. My wife tells me there are two of you: there's the one that's at work, which is the one everybody loves, and then you come home, and we get the leftovers. In some way she is correct. I give everything I have to the people that I come across because I never want them to feel the way I felt when I was in Indiana. I don't want them to ever feel that, no matter how busy I am at work. I never make them feel as if somebody else is more important to me than they are.

No matter what I'm going through at home, I try not to let that affect how I care for them or how I interact with them. I worked at the residency clinic for four years as an LPN, and then, when I got my RN degree, I started working on the neuroscience and spine unit at the St. Cloud Hospital. I got the DAISY Award last year. It's an award that's given out once a year. I got it because a family

I took care of recommended me for that award. I don't know which family it was—I tried to think about who it was to say thank you.

Besides having my kids, that is one of the most fulfilling things in my life other than meeting my wife. I go into work every day and try to make a difference in somebody's life—a positive difference in somebody's life. That's what I try to leave in my community. I will be a good ambassador for anybody whether they're black or white—you will have a good impression of people in general.

greencardvoices.org/speakers/itoro-emmanuel

EUROPE

Cork, Ireland

Elaine Black

From: Cork, Ireland
Current City: St. Paul, MN

Field: Microbiology

> "I FEEL ENERGIZED BY THE WORK I DO . . . I FEEL I CAN CONTRIBUTE AS A MENTOR FOR OTHER FEMALE SCIENTISTS AND TRY TO HELP GRADUATE STUDENTS UNDERSTAND HOW TO MOVE FROM ACADEMICS TO INDUSTRY."

I was born in Cork City, Ireland, and grew up in Castletownbere, a remote fishing village of about eight hundred people. Castletownbere was an idyllic and beautiful place to grow up in. It sits on a rocky yet green peninsula that juts out into the Atlantic Ocean. My father was a house painter who died at an early age, leaving my mother to raise three little girls ages five to eight. Despite the obvious financial and emotional hardship of this, my mother ensured that we had a lovely childhood. We spent time on the rocky beaches, lots of outdoor play with the wild neighbor kids.

We also got to enjoy our granddad's farm before he eventually moved in with us. I have great memories of the farm: looking for lost calves at Easter time, bringing the bull "Curly Wee" to the well, and playing in the hay shed. My mother was and continues to be an important role model in my life. She received support from Ireland's Department of Employment Affairs and Social Protection, and she also worked numerous part-time and full-time jobs to supplement our income. Ireland has an impressive education system with access to governmental grants to ensure everyone has a chance for third-level education. Both of my sisters and I received great educations in nursing, science, and architecture.

As with many small villages in Ireland, there is little to keep younger people there. The economy of the area is reliant on the fishing industry, and there are obvious highs and lows depending on weather and fish quotas. Our mother encouraged the ambitions of my sisters and me, even though it meant we would likely leave the area. She always encouraged us to do our best, and I believe we have made her proud. My youngest sister and I moved to Cork to go to college. My oldest sister attended nursing school in Dublin and eventually emigrated to the Middle East.

While my childhood was a happy one, my teen years were not. Secondary school was a struggle with bullying and small-mindedness often prevailing. It was a relief to go to college, reinvent myself, and spread my wings. I embraced college life and the opportunities it gave me. I completed a bachelor's in food science and was encouraged by my professors to continue in academia and immediately earned my PhD. During my studies, I took any opportunity to travel, attending many conferences in Europe and receiving a scholarship to study in Belfast, Northern Ireland, and the University of Delaware. My six-month stay in Delaware was my first experience in the United States, and I returned after my PhD to work for two years at the university as a postdoc.

Moving to the United States of America

I arrived with two suitcases and found my first solo apartment. Living alone was very new for me. In Ireland, where it's expensive to live, students and young professionals often live in a community—an apartment with three or four people or a house with four or five. My very Irish social nature found this new arrangement difficult to adjust to. Luckily my time in Delaware was an amazing experience because I was quickly consumed by the International Students' Club. My small apartment was often full of people from all around the world—and two Russian guys became my best friends.

Although there is a strong connection between Ireland and the United States, when I arrived in Delaware, it didn't feel like I had emigrated. I never had a desire to pursue the "American Dream" that had been portrayed by generations of immigrants before me. I was simply gaining work experience in my field with a plan to return home in two years to teach or continue research. I had never expected to make my life in the United States, but the recession hit, and there were no jobs in Ireland. My connections in Ireland advised me to stay in the United States. Here was the turning point at which I began to think of myself as an immigrant. I was no longer here for fun, adventure, and work experience—I was here for my new life.

STEM Career

In 2008, it was difficult to find a new position: universities were not taking chances on foreign candidates for professorships. I took a postdoc job at the University of Minnesota. It was very different from my experience in Delaware. Out east, I lived in a small college town with an amazing array of international folks. I moved to the Twin Cities in January 2009. I didn't know anyone, but luckily got connected with Irish American friends of my former professor in Delaware. They all but adopted me. They picked me up from the airport and welcomed me into their home until I got settled in my own place. They remain very much in my life and have become dear friends of my family in Ireland, staying with my mother when they visit Ireland. It remained a challenge to build a community of friends in the Twin Cities, but I persevered and found a lovely Irish expat community to ease the homesickness.

For the last nine years I have worked for Ecolab. I am eternally grateful for my supportive Australian boss who quickly helped me secure my green card. The stability of not having to worry about my next visa and going back to Ireland to search for employment has been immeasurable. I started as a microbiologist but have transitioned to a job in regulatory affairs. I work on advocacy and Environmental Protection Agency (EPA) and Federal Drug Administration (FDA) policy issues and represent Ecolab at trade associations. I feel energized by the work I do and the differences that a company like Ecolab can make in food safety, public health, and water security. I also feel I can contribute as a mentor for other female scientists and try to help gradute students understand how to move from academics to industry.

On a personal level my connection to Minnesota grows and grows. I married my husband, Joe, two years ago, and we have a sweet four-year-old boy. My in-laws are welcoming and warm, and their love and acceptance gives me a new sense of place here in Minnesota. I have made myself a promise to take my son to Ireland as often as I can so he can understand and fall in love with his heritage. It is important for us to also stay connected to the expats in the Twin Cities and the other kids with Irish parents. Having this little "half-Irish" boy in my life has made me more connected to my family at home in Ireland and keeps me grounded where I have made our other home.

greencardvoices.org/speakers/elaine-black

New Delhi, India

ASIA

Manish Shahdadpuri

From: New Delhi, India

Current City: Blaine, MN

Field: Information Technology

> "I CAME HERE IN THE US WITH JUST MY PARENTS' BLESSINGS, TWO SUITCASES FULL OF CLOTHES, A FEW SNACKS, AND A FEW DOLLARS I GOT FROM INDIA. THAT'S ALL I HAD. OVER THE YEARS THOSE TWO SUITCASES HAVE EXPANDED TO A HOUSE FULL OF THINGS. IT'S GROWING FOREVER."

I was born in New Delhi, the capital of India. All of my years in New Delhi were spent with my parents, family and friends. The life in India was pretty much stress-free throughout the years I was there—no obligations . . . I was a free bird who was used to getting whatever I need.

My mom and dad used to work as center government employees. They used to work hard to fulfill our needs and raise the two kids—me and my sister. I was born in a middle-class family, but as I progressed, as I grew, by the time I did my schooling, my parents graduated to be an upper-middle-class family. We used to get pretty much VIP treatment wherever we would go because of my dad's position and my mom's position. So, I never had any issues, you know, getting any stuff done back in India—going to various places, living in the resorts, spending good time with family. I spent a lot of time playing and watching cricket and a lot of other sports. I played sports and played on the streets. I never got any professional training, of course, but I was involved. I pursued my hobbies in sports and in music. My dad used to describe me as a "jack of all trades but master of none" because I didn't specialize in any one thing, but I was good at everything. I used to explore everything. So life in India was pretty cool and relaxed. I spent a lot of time with friends and family, and I had fun all the years I was there.

Moving to the United States of America

My move to the US was really more of a casual decision. I started working around 1995, 1996, and I was fortunate to get an opportunity to travel abroad on a short assignment to the UK. In my second job, I was given a chance to travel to the US on a small assignment; then I went back. I was happy where I was working. I had no strong desire to come to the US on a work visa or anything. But then, around 1998, 1999, it became a trend where if you were a good, educated professional, you would want to travel abroad and go to the US because the US had lots of job opportunities at that point in time. A lot of my friends were grabbing those spots and just going to the US. So, I started contemplating if maybe I should try it out—explore a little more the professional work culture in the US, experience a different culture, go with my friends, and enjoy a life in the US for a few years.

With that in mind, I just casually started applying for jobs in Boston with the recruiting companies. My friends chose a particular recruitment company to come to the US, and my visa was approved by that company as well. So, I thought, "Yeah, might as well go there, spend some time in the US, enjoy life with my friends." I didn't have any financial obligations. My parents were pretty young, self-dependent—I had no hesitations. So, I thought I would give it a try, go to the US, spend some time there, experience it, and come back. I really loved the welcome and the culture, the people's hospitality, and the attitude a lot more in the US. I just thought it would provide me much better possibilities, much better experience, and opportunities to grow managerially and professionally, so that's why it was the US over any other country. Plus, there were a lot of IT jobs in the US that time especially.

I came to the US in the year 2000 and first landed in Chicago. From Chicago I took a shuttle to come to a small suburban area called Moline. That is where my parent company was. Three of my other friends, who came in the same company, came to Moline a week before I came in and joined the same company. When I arrived in Moline, I think it was the month of July. My friends and I used to live in a small apartment provided by the company—four of us together. We didn't have a car, so we used to walk to the office. It was a fifteen-minute walk. We used to feel so happy if somebody would offer us a ride from the apartment to office or back. I worked on the project in Moline for a couple of months, and then I was sent to a client as a consultant in Boston. I stayed in Boston for around three months, and I worked for a dotcom company as a consultant. At that time there was the dotcom bust, and the impact of that was that the company where I worked got shut down. So, then I had to come back to Moline. I spent another month in Moline before I got this opportunity in Minnesota, where I came in January of 2001. That was probably the coldest winter I have ever seen in the United States. I landed January 25 in Minneapolis, and I was freezing cold. That was a real shock to me to see all that snow the first time . . . piles of snow.

When I left India, the most difficult part for me was leaving my parents, even though they were pretty independent. But I'm very emotionally attached to them, so leaving them back there was a little difficult for me. Of course, I also had to leave the authentic Indian food, the home-cooking, and the outside food that we didn't get that frequently in the US. There were not a lot of restaurants around, so we used to cook—but we are not that great of cooks. Having to leave my friends and all the casual things we used to do back in India, when we came here we realized how much all these small things matter—the food, friends, and family. I still remember that calling back to India used to be very expensive. It was around fifty cents per minute I believe at that time. We had to dial the toll-free number and long passcodes, and wait—sometimes forever, hours—to get through the connection, and talk to my parents for a few minutes there. So that was very difficult. It was not as well connected as it is today. There was no technology to do Facetime or have video chats and things like that. That was something really difficult.

In India, I used to own a car myself, and I used to drive around. I never used to walk a lot. In fact, I took my friends all around in Delhi. But here, when we came, I didn't have a car. I used to be dependent on others for a couple of months initially. That was difficult. Learning the car wasn't difficult. Getting a license and getting the car was the difficult part. There was just one walkable restaurant, which was just the McDonalds. We used to walk there, and my friends and I were vegetarians on a few days—I'm a vegetarian on Monday, Tuesday, Thursday. So, we used to walk to McDonalds, stare at the menu for a few minutes to decide what we can eat, and eventually we would only order a veggie burger—which is just bread, lettuce, tomatoes, and cheese in there. We did that for quite a few days. We didn't have any other option. It was a little difficult initially to get used to the food or find what we liked, or try different cuisines. That was a challenge initially. Sometimes there were also problems understanding the American accent. I remember we went to a grocery store once and at the check-out counter the lady asked me, "Paper or plastic?" I heard it as "pay for your plastic." I said, "No! I'm not gonna pay for my plastic!" But slowly I got used to understanding the different accents that Americans use here.

I still remember the first weekend my friends and I had here. We were so excited to go out for the first time. We all got dressed up and called the taxi to go out. Where were we going? We were going to Walmart to do groceries. So excited! It was our first outing. It was interesting. In the end, I was living with my friends, four of us coming together from India, staying together in the apartment. It was a fun, good, enjoyable time. We still cherish those moments when we get together; we talk about those times. It was a good time we had, slowly getting used to the environment, the culture, professions, outside . . . everything. That was good.

STEM Career and Cricket

I came to Minnesota in January 2001. I joined the company SDRC. I've been with the same company for the last twelve years now. The company has changed hands a few times. A few years back it was part owned by Siemens. By now I am pretty much settled in Minnesota. I'm used to the cold here, and I don't dislike it anymore. In fact, I feel that in winter you get more time to bond with your family and to spend time with your kids. I feel Minnesota is a very nice place to raise a family. I don't think I'm going to move out from Minnesota. I went back to India and got married in 2002, and we started a family. In 2004, we bought a house. We have two kids. I had my daughter in 2004 and then my son in 2005. They're going to good schools. We are raising—I believe—law-abiding, good citizens now.

I started my master's degree when I got married and then completed it five years later with two kids. That was an achievement. Today at Siemens I manage a team of around thirty people distributed all across the world—India, the US, and a couple of other locations. That's a pretty challenging and interesting job. I work as a technical manager of certain big distributor teams and track various parts of the organization, including the customers and technical teams. It's a pretty interesting job. I'm happy where I'm working today.

I also get to continually pursue my passion of cricket—not only watching cricket, but also playing cricket. When I came here to Minnesota, I figured out there was a cricket league here. It's a nonprofit organization association. I joined that in 2002. I was also involved in founding a new team called Friends Cricket Club in 2003. We are one of the best teams in the Minnesota Cricket Association (MCA) now. If I was in India, I don't think I would have been able to continue my passion of playing cricket. Here in the US, I get time to pursue that passion, and I've been playing cricket for the last twelve years now. That team has been expanding and growing—it's one of the most solid teams in MCA.

The US is a country that embraces diverse cultures and diverse people from various parts of the world. I am trying my best to add more colors to the diversity of this country. Recently MCA started a program to train the youth in cricket. I was the head coach for Blaine Cricket of the Blaine City youth program last year. I was training the kids, and it felt really good to see the kids play cricket and get trained. The hope is that in a few years the kids and youth here will get an opportunity to get coached and trained, and cricket will get played at all levels as a mainstream sport here in the US.

Contributions

Over the years we have also started to organize and participate in a lot of community events. I pretty much organize a grand New Year party every year. We get a lot of people from the Indian community to come close and greet together and have fun at the same time. Also, my wife and I volunteer at various events at work and at school. My wife was the chair for the multicultural event at my daughter's school. I have volunteered to represent India at the multicultural nights to share the knowledge, the information, the culture, and the traditions of India with people here in America. Also, we do our share of charities. I support a couple of charities. We also do more than our part of spending to keep the economy of the US going.

I think the US is a very good place to raise a family where we can give our kids the best of both worlds—a blend of the American culture, the western culture, and still stay close to our grassroots. For example, my daughter, on one hand, is now learning jazz, ballet, and tap dance, and on the other hand, she is going to Indian classical dance classes as well. She's learning piano, but at the same time she's attending classical vocal music lessons. My son, he's very passionate about playing football and basketball, but at the same time he loves the sport of cricket and wants to get trained in that sport as well.

Over the period of years, we have built very good relations with a lot of people here. I have a big circle of very, very good, trustworthy friends. I can really depend on them. We share very good relations with our neighbors. We do get-togethers at times with friends and share traditions with each other. Sometimes during an Indian festival, we will take sweets to their houses and tell them what this festival is about. And during the American festivals, they will come to our house and share their joy. At my work as well, on Diwali, we organize an event where we share the joy of the festivities with our American colleagues as well.

I came here in the US with just my parents' blessings, two suitcases full of clothes, a few snacks, and a few dollars I got from India. That's all I had. Then I moved from Moline to Boston, Boston back to Moline, and then Moline to Minnesota. That's all I used to carry—two suitcases with me. Over the years those two suitcases have expanded to a house full of things, and it's growing forever.

greencardvoices.org/speakers/manish-shahdadpuri

NORTH AMERICA

Reynosa, Mexico

Karina Boos

From: Reynosa, Mexico

Field: Product Engineering

Current City: Tonka Bay, MN

> "WHEN I FIRST DECIDED TO GO TO COLLEGE, MY DAD ACTUALLY DID NOT WANT ME TO GO TO COLLEGE. HE USED TO SAY, 'WOMEN USUALLY JUST MARRY, HAVE KIDS, AND STAY HOME.' I THINK THIS WAS PART OF THE CULTURE IN MEXICO."

I was born and raised in Reynosa, located in the state of Tamaulipas, Mexico. It is a border town with Texas. Growing up, I remember that Reynosa was a very small and tranquil city. I had a good childhood there. My parents, Benito and Socorro, migrated to Reynosa from Tampico, which is located in the southern part of Tamaulipas. They moved mainly to work there. My dad was a labor worker for Pemex, the national oil company in Mexico. It was a good job for him. My mom was a stay-at-home mom. My fond memories from my childhood are when my dad would take us—me, my older sister Laura, and my mom—to the movies and to eat dinner, usually seafood. Every summer we traveled to Tampico to visit family and go to the beach.

STEM Education

When I was in high school back in Reynosa, I was always interested in STEM. I went to a technical school, CBTIS 7, where you can get a technical degree beside your high school diploma, so I graduated with a lab technical degree. After completing high school, I did not actually want to pursue a career in chemistry because there were not a lot of opportunities to work in that field. For college, I was inspired to study engineering due to the boom of manufacturing plants in the border area in the nineties after the NAFTA agreement between the United States, Mexico, and Canada, so I got my BS in industrial engineering in 1998 from the Autonomous University of Tamaulipas, UAT.

When I first decided to go to college, my dad actually did not want me to go to college. He used to say, "Women usually just marry, have kids, and stay home." I think this was part of the culture in Mexico. So there were some things that I had to overcome. Money was an issue. My mom went back to work after being a stay-at-home mom for many years. She had an accounting degree, but no experience, so she worked as a housekeeper for the Pemex Hospital. I am grateful that she gave me the opportunity to go to college. I remember that a few times we had to borrow money from my Aunt Elvia to pay for my tuition, but then I got a part-time job. I was also able to start studying English as a second language. I had really good friends in college—Maribel, Gaby, Yolanda, and Brenda. We always helped and supported each other. I still keep in touch with them.

My first job as an industrial engineer was with Delnosa Delphi Electronics in Reynosa working on the EV1 Engineering team. EV1 was the first electric car from General Motors (GM). We manufactured the charger and power unit. My job was to create work instructions for the manufacturing lines, cycle timeline balance, and continuous improvement. GM killed this project after a few years— the battery technology for electric cars was not fully developed at that time. From Delphi I moved to another manufacturing plant, Magnetek. They made lighting ballasts. Working there as a manufacturing engineer, I had the opportunity to get trained and certified as a Six Sigma Green Belt. Unfortunately, this company had to relocate the facility to another city.

Moving to the United States of America

Then I joined Eaton Corporation. I was part of a transfer plant team. We were transferring one of the Eaton Hydraulics facilities from Carol Stream, Illinois, to Reynosa, Mexico. So I came to the United States for temporary work in the year 2000, and I was traveling back and forth between Reynosa and Carol Stream doing manufacturing engineering work. While working in Reynosa, I volunteered to take the lead to manage prototype orders and became more familiar with new product development and design. Then there was an opportunity at the Eaton headquarters in Eden Prairie, Minnesota, for a product engineer position. I was very interested in applying. The hiring manager was very familiar with my experience and my work, and she advocated with HR to consider and justify an international relocation. After a few rounds of interviews, I got the job. Eaton paid for my work visa to come here to Minnesota. It was a very long process to certify my degree in the US and to fulfill all the work visa requirements. I was really excited to come to work in the US. It was a really good opportunity for my career and my development. It was a new adventure.

I moved to Minnesota in November. It was a really cold winter, and it was really hard to be very far away from my family. When I first came for the transfer of the manufacturing plant, I was here with a team of co-workers, but this time I was all by myself. I didn't know anyone. My parents and my sister were still back home in Mexico, so it was a really difficult time for me being alone away from home. I had to live in an apartment by myself, and I only had a small TV, an air mattress, two chairs, and a bag full of dreams. At work I kind of knew some of the coworkers from the engineering team. They made me feel welcome. They used to tell me to be very careful driving in the snow, to wear good shoes for the snow, and how to be safe out there during the cold winters in Minnesota.

When I first came to Minnesota, I told my mom that I was going to be here for only three years working under my temporal H1-B visa, but actually I formed a family here. I got my green card through marriage. I married an American. He also works for Eaton as a Field Service Engineer. Dan, my husband, is Caucasian. He is from Wisconsin. We are blessed with two boys: Ryan, 15 and Erik, 11. Our little family is biracial and bicultural. Ryan and Erik attend the Minnetonka School District which I love because they offer Spanish immersion programs. They both play American football and basketball.

So after many years, I consider Minnesota my new home. I really like living here because you get to experience all the seasons. I love fall and winter. I love playing outside with my kids and going sledding. My parents, my sister, my sister's family—Cesar, Alex, Edgar, and Melanie—all still live back in Mexico. I take my husband and my boys to Mexico at least once a year to visit family, and we go to the beach as my dad used to take me every summer. I also have a great American family in Wisconsin. I have seven brothers-in-law, and now I celebrate new family traditions like Thanksgiving. We also cheer for the Green Bay Packers!

STEM Career

When I started working as a Product Engineering here in Minnesota, it was a bit of a challenge for me because I had to do mechanical engineering work, which I didn't go to school for. Because of this, I had to overcome a learning curve. Most of my experience was in the administration of the manufacturing process and operations when I had worked for manufacturing plants. For Product Engineering I had to learn different programs and engineering calculations. Fortunately, I had good coworkers that took the time to train me—I actually had formal training for Pro-E to design 3D models. I went back to school and took some classes and certifications to be able to do a better job. As well, I was able to apply my Manufacturing Engineering experience to New Product Designs. For the Eaton's Power and Motion Controls product lines, I have released more than 100 Custom Manifold designs for OEM customers like John Deere and CNH for the Off-Highway mobile equipment market. I also released two catalog items for screws in the cartridge valves product line.

At Eaton we recently started a new program for inclusion and diversity. We have Inclusion Employee Resource Groups ("iERGs"). These groups are voluntary employee groups based on shared backgrounds or life experiences. I'm a volunteer and Site Lead for the Vamos ERG, a Hispanic and Latino group, I'm really proud of the work we are doing, Eaton is a global company. We have sites in Mexico, Brazil, Puerto Rico, and many other Latin America locations. The goal of the iERG is to attract, develop, and retain Hispanics and Latinos. We also work to have representation and different perspectives to solve problems and create solutions in the global economy. We have a partnership and collaboration with the Society of Hispanics Professional Engineers ("SHPE") Twin Cities, and last year we had our first win—Eaton recruited a mechanical engineer through this networking opportunity. I am very proud of the person who got hired. He is from Peru, and he is actually a DACA recipient. He is really smart, an excellent person, and a great asset to the company. I was really happy to hear that my company is giving opportunities to Dreamers.

What is special about STEM here in Minnesota is that we have a lot of big corporations headquartered like 3M, General Mills, Target, Boston Scientific. There are a lot of opportunities for STEM careers and to invest in the development of the creation of new technologies. It is a very competitive job market too, and for me, as an immigrant to be able to compete with everyone else is important. I'm currently pursuing Masters in Systems Engineering at the St. Thomas University. It was really difficult for me to go back to school after many years. Actually being in school studying in a different language, doing all the essays and requirements . . . it is very different than when I went to school back in Mexico. I tell everybody that we should keep working everyday to learn. In engineering there are always new things coming out that you need to keep up with, so that you can be competitive.

greencardvoices.org/speakers/karina-boos

AFRICA

Lagos, Nigeria

Sampson Abiye Linus

From: Lagos, Nigeria
Current City: Woodbury, MN

Field: Chemistry & Technology Management

> "I GREW UP FASCINATED BY AMERICAN CULTURE. SCOOBY-DOO—ALL THE SHOWS THEY HAVE HERE—WE WATCHED THEM. I THINK THAT DREW ME TO THE AMERICAN LIFESTYLE . . . [AND] CULTURE . . . I WAS FASCINATED. WHEN THE OPPORTUNITY CAME, I SAW MYSELF LEANING TOWARDS FINDING A LIFE HERE."

I was born in Lagos and grew up in Lagos for most of my life. Lagos is in the Western part of Nigeria, it's typically similar weather to California and close to the beach. I grew up in a middle-class family. My dad worked for NCR, and my mother owned her own business, so I had a good city life. We lived in the suburbs, but I was able to go enjoy some of the city life in Lagos. From time to time during vacations, we'd go to our hometown in Rivers State to keep in touch with my culture and tradition and interact with my cousins over there. My life was kind of a life in-between both the suburb and rural; I also had a lot of contact with my family and that impacted my development and growth as a person. It gave me a balance of both city culture and exposed me to more European and American culture versus just rural traditional culture as is the case for some people in Nigeria.

I grew up fascinated by American culture. Surprisingly, when I interact with people born here, we find that we both watched similar shows growing up—it's amazing. Scooby-Doo—all the shows they have here in the US—we watched them overseas. I think that drew me to the American lifestyle and American culture. I was fascinated, and so when the opportunity came, I saw myself leaning towards finding a life here, developing my career, and building a future here.

Moving to the United States of America

The Diversity Visa Lottery—the "DV visa," as it's known—is a visa program run by the US government and the US Department of State through embassies. It's an opportunity given to immigrants from all over the world to be able to work and live in the United States. It also opens up opportunities or paths for them to become United States citizens after five years. This program is designed to select the best possible people who are able to come to the United States to find a life as well as those who are able to assimilate into the culture and values of the American dream.

My sister brought the forms, she applied the year prior. I applied the first time, and I got notified after about a year that I won. It was my first attempt. I reached out to the embassy, and they notified me they will send more information after a certain time, so I had to go through a lot of processes prior to my interview. We had biometric tests done through the FBI and by the local police. Background checks were done; legal checks were done. Once we passed through those tests and were successful, we were put to the next phase, which is the medical and vaccination phase. We were tested for all kinds of stuff and then vaccinated also in preparation for the interview to make sure we're compliant. You also get a WHO clearance, which is a card from WHO saying that we've met all the medical/vaccination requirements. At that point they request a letter from a prominent person, usually a chief or a judge, that just says if I am a person of good character. Once that's presented to the embassy with all the documentation, we pay a $350 dollars just for the interview—this excludes all the costs for the medical and biometrics. At that point they schedule a date for your interview. I think my group was a group of seven that came in for the interview, and out of the group of seven, I was the only one that got the visa. I consider myself very lucky.

Unlike most other visa programs in the US, coming through DV visa requires a lot of financial output. Looking at the value of the dollar then was about 250 of my currency, the naira—so, of course, it's your life savings, what you've earned over the years. It was a lot of money spent. Then you have to talk about buying a ticket, and, of course, you have to show a deposit in your bank account to show the embassy that you're able to live on your own for at least six months when you get here so that you will not become a public charge. Coming under DV requires a lot of financial investment in yourself, and if anybody would look at that and think maybe they should invest this money in businesses here, they wouldn't really want to come. It takes a lot to go through that process.

Typically, most people stay with relatives when they come to the US. My closest relative to this place was in Boston. I'd done my research through the embassy to see what was the best place to go to and potentially get a career and raise a family because the stereotype of cities portrayed in Hollywood movies are huge overseas. Minnesota ranked top of the states. I think it was ranked number two at that time, but I didn't know anybody in Minnesota. I do have a cousin here, which I found out later on, but at that time, I didn't know anybody here. I came to Minnesota, and I stayed with distant relatives of a friend of the family when I got here. My plan was to come in, stay a short time, and find my life. It was a good challenge for me. It was a little scary leaving: it's like leaving everything you know behind. But I think that's one of the points of being an immigrant, the risk. I was ready for that.

The good thing about coming with a DV is you have your visa and your green card in your hand. Coming into the US was pleasant—the Immigration and Customs officers were very nice. Actually, one of them gave me a free hat on my arrival at MSP airport which read "Welcome to America," and that warmed me up and just told me people are more welcoming here. I came in October, so it wasn't too cold but a little chilly. I remember that it was a sunny day. I went to the airport tarmac, got a taxi, had the address, and went straight there.

The first week was busy for me. I had to go apply for a driver's license and testing. I had to go to the Social Security office with my forms and notify the INS that I arrived with paperwork. There was a lot of movement around for me, going to the county office to put in my resume for job applications. My first week was very busy. I couldn't drive and didn't have a car, so I spent a lot of money on taxis. It was fun. I learned to take the bus too, so that was good.

STEM Career

My process of assimilation to prepare myself for my journey was to visit the county and gather information about job postings and opportunities. Dakota County notified me they had a program called the Diversity Training Program. It's a program sponsored by the state, which they get funded for, where they had immigrants from different parts of the world here in Minnesota come into the county, get trained, and learn how things are done in the county, especially those with science backgrounds. Being a chemist, I worked with other chemists, microbiologists, and IT people who were from different parts of the country.

I was given the opportunity to work with a very intelligent lady called Maria Endrody. She has a PhD in chemistry, so it inspired me to learn a lot from her. She shared all this knowledge and wealth from the work she's done in Hungary, and I shared my work from Nigeria. We used our knowledge to help accomplish projects. This helped me understand the power of diversity and learning from other groups. What that program did was open up the opportunity for me to learn about business practices, regulations, and writing for reporting of the standards of the county and the state. That prepared and helped me in the US. I unfortunately learned that the program is no longer available in the county, but I think that's one program that would benefit immigrants if they brought it back. It did help me a lot to assimilate and learn and accelerate my learning of processes in the States.

I would say I was self-motivated to go into the field of chemistry. My initial plan was to be a medical doctor. It's a competitive world. Career option two was chemistry. I decided to go after that second option. I was self-motivated to go into that field and come to the States, but it's a whole different game here. It was very competitive when I got here. I got offered pay that was lower than I expected. Back home I was a quality control manager, so I understood the importance of having the lab background, knowledge, and management skills. I was a member of the Nigerian Chemical Society, and also a member of the Institute of Chartered Chemists of Nigeria. I learned to interact with administrators and managers in the STEM field, and they told me to rise up in this field, you have to not just a knowledge of the chemistry or engineering—you must also be able to have some business knowledge to be able to get up the ladder and lead teams. I had that in the back of my mind when I came to the States.

When I got here, I looked at the job market and saw the competitiveness. I met a couple of immigrants who were in the field, and I saw that they did a great job, but there was a bit of a challenge in terms of rising up the ranks. I decided to get out of my field for a second and go into the business field and learn. I decided to get my MBA and learn Lean Six Sigma, and today I am a black belt in this field. I gained additional skills outside of STEM for a bit. I am fortunate to find myself with a very good company that gave me the opportunity to grow up the ladder. And the combination of the skills in chemistry, business, Lean Six Sigma, project management, and team building accelerated me and my career in the company. I would say my journey's been a little bit of a loop, but at the end of the day, the loop ended up taking me faster to my goal.

I would say being an immigrant is very tough. When I came to the States, I found it interesting what the average chemist was paid here. In Nigeria, the chemist is ranked like the medical doctor, like a general practitioner. I made a ton of money back home. Coming to the States, it was less because we were chartered chemists back home, but here you're just a guy in the lab. I learned to be able to continue my journey aspiring to leadership.

I would say I aspire to be better. There is a lot of competition in STEM, but there's still also a high demand. I think to be successful in STEM you must be flexible and be able to move. I have moved twice in my career—to California and back to Minnesota. I think being able to diversify your skills in STEM and possibly further your career by getting a PhD, getting an MBA and expanding your sphere of knowledge goes a long way. There are a lot of opportunities, scholarships, and grants that can be used. Some companies pay part of the tuition, so it helps. I would say, as an immigrant, explore avenues where you can get resources and also possible funding to support your career path. I think it does help a lot because being an immigrant is difficult, but with the opportunities and determination, I think the American Dream is not far away—it's attainable if you put your heart to it.

I've been married for eleven years. I've known my beautiful wife for twelve and a half years. We have three lovely kids, three boys: Joshua, Caleb, and Daniel. They're my joy. My wife Darlene was born and raised in the US. She is one special person to me because she is open to my culture. She made America my home for me. She surprisingly learns my culture—she cooks our meals, wears our outfits, and identifies with my family. Teaching my boys also to appreciate that culture goes a long way, and it kind of steers away from the stereotypes. We're the same people, and we just have to understand each other.

Contributions and Impact

For me, identifying with and understanding what's happening in my community is important. I am part of my company's diversity drive and part of the employee group. In addition, I am also part of outreach to the community that works with a group that helps African American men in society who are reintegrating back into society after incarceration. I also work with a group that does cultural diversity, what we call the "Cultural Lab." We've been able to work with several groups. I think the first was a Latino group, and now the second group was in the Ethiopian community. I was part of that discussion to help develop ways of helping them career-wise, driving cultural ethnic blending into society. Being part of these conversations in my community has opened my eyes to the power of diversity and also the importance of having that common voice in society.

I love fishing and learned a lot about it from my late grandfather. Fishing in Africa is different from fishing here. In Africa, we're taught to fish with nets—we knit the bait, which takes a long time and is very tedious, and then go for a big catch. Coming to the States, I learned how to fish with a reel and fishing pole. You catch one fish at a time, which is a big difference, but I love fishing. My wife's dad is great at fishing. He owned his own boat when we were in California. Whenever he comes here to visit, we always go fishing. That's our thing. I love soccer, of course. I played soccer when I was home, played for my college, so I love to watch soccer now. I don't play actively anymore, but I am still a great enthusiast of soccer. I play ping pong a lot, and I've surprisingly become a very big football fan. I love football. It reminds me a little bit of rugby back home. My friends tell me, "If you were born here, you'd be a football player." I don't know how true that is, but it's fun. I'm a big Vikings fan. Other hobbies I have include writing music and poetry. I actually won a competition back in 2005, Editor's Choice Award. The title of my poem was "The Dreaming Immigrant."

greencardvoices.org/speakers/sampson-abiye-linus

NORTH AMERICA

Tegucigalpa,
Honduras

Valerie Ponce

From: Tegucigalpa, Honduras　　　　　　　**Field:** Clinical Research
Current City: Rochester, MN

> "IT WAS THE END OF 2013, DECEMBER 28TH TO BE EXACT, WHEN I BOARDED THE PLANE IN TEGUCIGALPA AND WAS ON MY WAY TO THE UNITED STATES. I REMEMBER CRYING A LOT BECAUSE IT WAS THE FIRST TIME I WOULD BE AWAY FROM MY PARENTS AND HOME."

I was born in Tegucigalpa, the capital city of Honduras. It is not the biggest city in the world, but it's very populated. It has a great climate. The one thing I will always remember is that growing up there was this store that had really loud music every Saturday and Sunday. I feel that this describes Honduras: very happy and lively—music everywhere. It's a place where people are warm and very welcoming.

My childhood was surrounded by loving people, family, and friends. We just went out, enjoyed the sun, and went around exploring Honduras since there are a lot of places you can visit, such as parks and museums. I was born into a family of four siblings, so I have three sisters. We grew up in what I consider a normal childhood—just playing, going to school, and doing kid things. I was very fortunate to go to a private school. That is where I learned how to speak English. I can't really remember anything remarkable of my childhood. I can just say it was pretty happy. My family had some struggles, but I think everyone does. My parents did a great job raising us.

Honduras is known—or has recently been known—for its violence, for not having enough jobs, so, consequently, people are migrating in increasing numbers. In the world's eyes, it is seen as this unlivable place, but to me it was a happy and vibrant place. I recognized that we had a lot of obstacles, but it was also a place where it was always warm, people were always welcoming, and the music was always playing.

Honduras is considered a developing nation. In 1998 there was a hurricane, and it really set back our development. During the hurricane, we continued our life as normal. All I remember about the hurricane is that it rained a lot. Natural disasters like this one and political instability are what keeps a beautiful country with amazing people from moving forward. Hondurans want better things, and we are hardworking, but there are multiple things we have to overcome before we are able to progress.

Growing up, I was always interested in and wondered about the why and what of things—the why and the how of things work and why they work a certain way. I always thought that being a chemist was pretty cool. The idea of mixing reagents and chemicals and learning the periodic table . . . I thought that was the greatest thing. You can say that I was always curious and interested in science.

When I was in high school, chemistry was just the coolest thing in the world. When I was researching degrees, I came across biomedical sciences, which is basically a workload of chemistry, biology, and physics. I thought, "That's perfect. This is it—this is what I want to do." Once I saw what actually being a chemist entailed, I was like, "Okay, maybe this is not my thing." But then I went on and took more and more biology courses and thought, "Okay, yes, this explains why things happen to me." Human anatomy was probably the greatest course I ever took, and also physiology because it explained a lot of what questions I've always had.

My family is not all in science. I think my dad is the only one. He's really into numbers, really likes math. So I think he would be the one who inspired me to pursue a science degree. He always wanted one of his daughters to become a civil construction engineer, which is what he does, but my sisters had different interests. I did not want to be an engineer because although I do like numbers, I'm not the greatest with them. I think it was mainly being exposed to science in high school and having the little influence of my dad that made me go into a STEM degree.

Moving to the United States of America

My school was very into the studying abroad idea for your college degree. My two oldest sisters studied abroad in the US, so this idea was always in the back of my mind. I initially wanted to study in Mexico or Honduras, but it just ended up not happening. So then I applied to schools in the US, focusing on schools that had science degrees. Studying in the US had its advantages because it is close to Honduras, and we are very familiar with the culture. We watched a lot of American TV and movies. When it came time to apply, I was like, "Yeah, I'm going to go to the US." When I studied abroad, I wanted to be somewhere very different than what I was comfortable with. Honduras has a tropical climate—it's either hot or a little chilly, but it's always nice, both in the drought and rainy seasons. I wanted to experience something different. When I was applying for schools, I applied mostly in the Northern US, and that is how I eventually came to Minnesota.

I remember packing most of my clothes. I was ready to leave what I had known all my life for this new experience. One of my sisters was still studying in Minnesota, so I was going to live with her, and this made the move less scary. It was the end of 2013, December 28th to be exact, when I boarded the plane in Tegucigalpa and was on my way to the United States. I remember crying a lot because it was the first time I would be away from my parents and home. One of my sister's friends came home with her that Christmas, so I ended up flying out with him. He helped me through the entire journey. We went through customs together. He bought me food. He tried to be supportive and said things like, "Just feel better! Don't cry—I know this is hard."

I arrived in Minneapolis on December 28th around midnight, and then I just waited for my sister at the airport because we boarded different flights. We have family and friends here, and they helped me get settled in and get ready for school. Also, spring is not the greatest time to start school in Minnesota, but that's what I decided to do. We got so many winter clothes, which I've never owned because in Honduras you don't need puffy jackets. We got scarves, gloves, and boots—the whole thing. I flew in through the Atlanta airport. From Atlanta it was a straight flight to Minneapolis. We have family friends who live in Minneapolis, so I spent the first week with them, and then I went to St. Cloud. We drove there. It is an hour north from the city. My sister and I were going to live in an apartment there, and that's where we lived for the first two years of my college experience.

I was moving to a different country and didn't know anyone. I was lucky to have my sister, but starting school is a completely new and different territory. They have an international student orientation at St. Cloud State University, and that's where I got to meet all incoming international students. It was like starting your first day at a new school and you're just like, "Ugh! Where am I? Who do I talk to? Why aren't people talking to me?" But this time you're in a different country . . . now you have to meet other people . . . and you have to be open to know a lot of people. My first few weeks were basically just trying to get settled in.

To me, the winter was just so shocking. I was just like, "This is really, really different." I definitely got what I wanted. My first few weeks were just getting to know people from all over the world. Culturally, it wasn't a really big transition just because Honduras is exposed to a lot of the American culture. I kind of knew what being in America was, like from what I saw in the movies, which is not totally accurate but it definitely gives you some perspective. My first few years, I befriended people from different countries from all over the world because they understood where I was coming from. I related to them better because I was going through the same or similar experiences as they were.

The US is great. There are so many different cultures that I was never exposed to growing up. Obviously food was one that I miss a lot. I don't really see how much I left behind, other than my family and friends, but family would be the biggest one. When I came to the US, my niece was a newborn—she was three months old when I left—and now I barely get to see her. It's like, "Oh, I hope she remembers me, but probably she does not." I only went back home for winter breaks, so she saw me once a year. I think that was the hardest thing, leaving my family behind and starting a life without them. A lot of the communication we do is through Facetime or WhatsApp. Now that she's six, she doesn't have the greatest attention span, so I talk to her briefly, and then she's like, "I just want to go play." Even though you want to talk to her for an hour straight, she's like, "Well, I'm done," and leaves.

STEM Career

After I graduated, it was time to apply for jobs to get experience. Lab work is fun, but I did not see myself working in a lab, so when I was applying for jobs, I was looking into science careers that focused on research. That's how I ended up in clinical research, and I really enjoy it—this is a great career. There're a lot of aspects of it. During school, you learn the steps of a trial and all the different stages, but don't actually know the work behind it. When you work in this field, you understand the way things work and know how much work is behind on a trial.

I think there is hardship because, with an OPT permit, employers don't want to hire you since they know you have just a limited time to work. I think that was the greatest struggle I faced when I was looking for a job. They ask, "Are you gonna need sponsorship?" Answering this is tough because it's like, "Do I need that? Do I need sponsorship? I just need some experience."

Minnesota is a great place to be in the health care field. There are so many industries and so many companies focused on health care. That's the biggest thing that I've learned. There's so much opportunity, and there will be opportunities that you wouldn't necessarily get in different places. For instance, in Honduras there are not that many research device companies. I mean, there are hospitals, but they don't really focus on research because it's just more of health care, like taking care of the patients. I think the biggest thing that I've learned is that there's just so many different sides of health care, not just going to medical school. There are things that need to happen so that doctors can have all their supplies and knowledge to treat a patient. There're a lot of people behind it—behind all the practices—and behind all the new instruments, devices, and techniques. There are just so many people that work behind it, and I think that's the biggest thing I learned here.

I think I live a pretty ordinary life, very relaxed and calm. I go to work, and then I come to the cities because it's just an hour away. I just enjoy spending time doing young people stuff. I enjoy eating out—that's probably the thing I do the most, just eat out. I enjoy exploring new things and trying new cuisines, so basically I eat a lot. I tried taking on new hobbies. I tried learning how to play tennis because I am not sporty at all. I took pottery classes. I just tried to become a better version of myself, a different version of myself that's not through work.

Contributions

I think my community has changed since I left college, because my college was very focused on just people from different parts of the world. Now that I'm in the work environment, I am more involved with American citizens, so I bring a new point of view and a new perspective because they see things one way, and I see it a different way. I befriend people. People say that I'm very nice, and they like to say that I'm social, but I don't know about that. I just talk to people, make them feel welcome.

Sometimes I volunteer for a clinic, and here I interpret for them since Spanish is my first language. There's a large Hispanic community in the surrounding areas of where I live. I try to be a good citizen and make small contributions to others by being caring and taking care of people, and just being aware of what is going on around me.

greencardvoices.org/speakers/valerie-ponce

EUROPE

Kukës, Albania

Ingrit Tota

From: Kukës, Albania   **Field:** Computer Science
Current City: St. Paul, MN

> "WE DIDN'T KNOW MUCH ABOUT MINNESOTA. I LIKE TO SAY THAT MINNESOTA CHOSE ME IN A WAY. [. . .] SOMETIMES I WORRY BECAUSE WHAT IF SOMETHING HAPPENS AT HOME? NOT ONLY I'M AN OCEAN AWAY BUT IMMIGRATION-WISE, DO I HAVE EVERYTHING IN ORDER? I THINK THAT'S BEEN THE HARDEST THING."

I was born in Kukës, which is a small town in northeast Albania. Many people here in Minnesota don't know where Albania is, and I have to explain where it is. It's a small country between Italy and Greece, a little bit north of Greece. My memories from childhood are great—doing homework most of the time and being a happy kid. I was born in a family of high school teachers and college professors. I have one younger brother. I lived with my brother and my parents and grandparents until I moved out when I went off to college. I had a very normal childhood. The earliest memories I have are around books just because everyone in my family was dealing with books. I think that has partially influenced my career choices and academic choices because I've been very driven to pursue an education. My parents pushed me to work hard.

Education and Study Abroad

When I was in my third year of high school, I applied for an exchange program for high school students. The program was originally started by Peace Corps volunteers who were working in Albania. They wanted some students to come to Minnesota and learn about the environment and bring their knowledge back home to try to raise awareness about environmental issues among the young people. I was one of the ten kids chosen to come here. We didn't know at the time where we would go. We didn't know much about Minnesota, so I'd like to say that Minnesota chose me in a way. That was just the beginning, and the rest of it is something that I had never thought in my life would happen. I didn't know I would end up coming back here to study, to live and to establish a future.

Everything started when I was a summer exchange student in high school. After the program ended, I went back to Albania to finish my senior year of high school. At the end of my senior year, I was ranked second-best student among all seniors by the national Ministry of Education. I also competed nationally in a physics contest and was ranked first. In the same contest in biology I ranked third. The results of this hard work partially contributed to boosting my self-confidence, and this was very helpful when I went to college.

After I graduated high school, I went to the American University in Bulgaria, and after my first semester there, I was part of another exchange program called "Work and Travel," which allowed college students from other countries to come and work during the summer in the United States. With the money they make, they travel and experience the culture and so on. I asked one of the organizers of my high school exchange program, Amy Fladeboe, if she could find someone who would be willing to extend a job offer to me so I could become a part of this program. She immediately contacted her mom who owns a coffee shop in Willmar, Minnesota. She said, "I have a great news! You can come here, and you don't have to pay for rent. You can live with my parents, and you can work at their coffee shop and stay with us." It was very amazing, kind of unbelievable, so I went to Sofia, Bulgaria and then flew to Minnesota.

When I landed, I knew who I was going to meet—I had seen their faces. I had seen them before for just a couple days when I first came to Minnesota. From the very beginning, they were very welcoming and kind. They drove me to their home, and from that very day I became a part of the family. I had the basement all to myself. It was set up like an apartment, and they never left me alone, in a good way. We went to Grandma Mary and Grandpa Jack's—now I call them grandparents—for happy hour. They lived a few blocks away. They took me to their family reunion and introduced me to the extended family. We would celebrate holidays together, like the Fourth of July. By the end of the summer, they basically saw me as their own kid, and I saw them as my parents. It is remarkable to mention that there were absolutely no financial benefits on their side from supporting me, and kind of embracing me. There were also no legal obligations. It just felt like we bonded. Laure and Jack Swanson, my host parents, have kids, but they didn't have kids together, so in a way they saw me as their kid.

It was the lunch of my first Fourth of July here in America when they said, "Hey, have you considered going to school here?" And my answer was, "No, it's too much work, too much paperwork, too much money." I came to America with fifty dollars in my pockets. They said, "Well, let's see what we can do if you want to go to school here." They were the biggest support I've had because they helped me find a college where I could transfer my credits and not lose anything. They also talked to their relatives, Tim and Stacy Seeman, who lived in St. Cloud and told me, "Hey, if you want to go to St. Cloud State University, you can come and live with us. We will drive you to school and back. You will be one of us." They gave me a room. They gave me food every day. They made cookies for me when I had finals. It was just amazing. Everything was lined up perfectly. Both the Swansons and the Seemans became my biggest supporters for my student visa. I was the last person to say yes to transfer my credits to St. Cloud State, even though things were lined up and everyone was excited about it. It was scary. It was scary to change the course of my life in just one summer and pursue something that I didn't know much about. I spent a lot of days doing research to see if there was a path, a future, if I pursued this. Everything worked perfect and here I am.

Moving to the United States of America

There were quite a few things I gave up when I moved to Minnesota to attend college. Being close to my family is one of those things. My family was an ocean away, and that was tough, especially when you are nineteen years old. When you first move out of your parents' house, going miles and miles away isn't easy. I gave up that comfort level that I used to have. I also gave up the comfort that my friends gave me. I moved away from everyone, which was tough. I moved into a reality that I didn't know from up close. I had seen America in movies . . . I had read books . . . but nothing compares to the real life.

I wish I had done a better job at recording all the culture shocks I've been through while here because now I notice that when my parents from Albania come to visit me here or when I go back, I experience them. I'm very eager to record them, to share with others. However, there are still a few things that I remember that did strike me when I first got here. For starters, I remember not seeing too many people on the streets. Back in Europe, I think people are outside more. They walk from point A to point B, and there is public transportation, whereas here I would see mostly cars. I remember at one point we were driving with my host parents and I said, "Look, there's a person walking." That was not very common, so that for sure got my attention.

People were very nice in Minnesota. I have visited other states as well, but I think in Minnesota, this is very obvious. When I was visiting Yellowstone National Park last summer, you could tell if somebody was from the Midwest, or especially Minnesota, because when you cross paths with them and you have five people crossing this way and five people crossing that way, you'll hear twenty-five hellos. Everybody greets everyone. People were very welcoming here. I felt that from the very beginning. I was also surprised that most people, or a good majority of them have pets, like dogs, at home. Even in Minnesota I was really scared of dogs. If I saw a dog, I would run away. But everybody, every household in which I lived, had dogs. I got accustomed to them pretty fast, and now I'm a dog lover. There are also other small things that do get my attention, but I don't think I'll ever get answers—like why is a dime smaller than a nickel? I keep asking everyone about that.

STEM Education and Career

I think my job is a combination of what I generally liked in the fields of science when I was introduced to them. I really liked math when I was growing up, but at the same time I was introduced to the computer. I remember my dad brought a computer home and told me, "It's yours—do whatever you want." After having studied computer science, I can tell that most of it is applied mathematics and algorithms. From the limited research I've done, I know that when you are younger, it is easier to learn certain things. One example of this is languages. The same thing applies to things like science and computer science. I think I was eager to learn how the computer works, and I applied whatever I could to learn more about it. After I was graduating high school, I knew that I wanted to do something that combined both math and computers, and that's why I studied computer science. I also did a pre-med track in order to pursue an MD in the future. It was an unusual combination of majors and a lot of work, but I was able to graduate with a 4.0 GPA overall.

Getting here and studying computer science at St. Cloud State University was partially due to the help I got from my two host families. Amy Fladeboe, one of my host families' daughters, started all of this and believed in me more than I believed in myself. Once I started going to St. Cloud State, I saw many doors and many opportunities that were waiting for me. I was part of a long-term internship with a company that hired students and then had them work for bigger companies while they were going through college and studying computer science. Then upon graduation, the company I was working for offered me a full-time job. That's where I'm currently working. So one thing lead to another, which made me a software engineer full-time today. But from this journey, I want to highlight the help from my host family, the encouragement from my friends, referrals, the hard work I did in different positions that I had, and my motivation to take up new challenges that helped me go up level, by level.

As an immigrant, the biggest hardships I think have been in terms of making sure I am remaining on status and traveling very carefully. At different stages, you have to make sure that if you leave the United States, you can come back, and that you have the right paperwork. After you graduate, you are in limbo where you are here legally, but if you leave it's very undetermined if you can come back until your paperwork comes through. That has been hard because I was unable to go home for Christmas for six years to Albania. That was a little bit difficult from an immigration standpoint.

To be here legally is a great opportunity and a blessing I would say, but it is also hard. It is a lot of immigration paperwork and racing with time and making sure you are staying ahead of things. That can be at times a little bit challenging. I've had people—friends and family here in America—tell me at certain points, "Why don't you just go to the DMV and apply for citizenship? It's pretty easy. I'll pay for the fees, and I'll drive you there." But it's not as easy as that—it's a very difficult process. Sometimes I worry because what if something happens at home? Not only I'm an ocean away but immigration-wise, do I have everything in order? I think that's been the hardest thing.

I have not worked anywhere other than Minnesota. From what I've read, and from the people I've talked to, I can tell there is a lot of potential and current opportunities in STEM here in Minnesota. Many people indeed do seek out opportunities in Silicon Valley. However, Minnesota offers, if you look at the bigger picture, a lifestyle where the cost of living is lower, people are really nice (especially if you have connections here), people you relate to, and good opportunities to work and develop professionally. I think that as a software engineer, and other STEM fields as well, there are many opportunities.

The lessons I would share with people who are following this path are to work really hard, but not just work hard, work smart and focus on what you really love doing. If you do what you love and what you're passionate about, it doesn't feel like a job. It will be something that you're feeling good about . . . something that you're proud of. And if you do that, I don't think you will ever get tired.

When I'm not working, I am probably writing code again, either for work or for my personal projects. I also like to hike a lot and swim. Unfortunately, in Minnesota that is limited because we have a very long winter. However, I try to keep up with that and to work out a lot because they say sitting is the new smoking these days, especially with the kind of job I have—so I try to stay in shape as much as possible. I also try to hang out with friends, go bowling, and mini golfing. I love board games, and I try to have some game nights with friends every now. I also spend time with family whenever it's possible.

Contributions and Impact
When I was going to St. Cloud State University, I think I contributed to

Minnesota mainly academically. I tried to help as many other students as I could by being an assistant to professors and a tutor. I also did undergraduate research, which I think benefitted me personally and also STEM in general. Then, as a software engineer, I do work for a company, but the product that we work on is a product used here in the United States by our clients. Generally, I think what I do every day helps people and entities in general and other companies here in Minnesota and in the United States. I also try to work on open source projects in the computer science area. I try to contribute as much as I can to the open source community and to my home. I've been here for so long that, yes, Albania is my home and will always be, but Minnesota has been my second, current, and hopefully future home.

When I was graduating college two years ago, my host family had prepared a surprise for me. The surprise was to fly my parents to Minnesota to attend my graduation. They were trying to do this behind my back so everything would be a surprise. I hadn't seen them in a while, and I was not expecting them to be here. It was wonderful because we spent not just my graduation but also Christmas and New Year's together. One thing led to another, and now they have a visa, and they have been coming back. It's been great having them here in the summer for a little while. I just feel very blessed because now I have three families—one in Albania and two here. When I see all of them together, it feels great. I am very grateful for not only what opportunities college and the companies have given me but also the people in Minnesota who have invested in me and helped me find a home. I'll always be very grateful to all of them for that.

greencardvoices.org/speakers/ingrit-tota

SOUTH AMERICA

Medellin, Colombia

Fernán Jaramillo

From: Medellin, Colombia **Field:** Neuroscience
Current City: Northfield, MN

> "WHEN YOU LEAVE, YOU LEAVE HALF YOUR HEART AND MORE THAN HALF YOUR SOUL BEHIND. . . EVENTUALLY YOU DREAM IN ENGLISH, AND I LOVE ENGLISH. I LOVE ITS POWERS AND POTENTIAL TO DO CERTAIN THINGS THAT DO NOT WORK SOMETIMES AS WELL IN SPANISH . . . IT GOES THROUGH A DIFFERENT PATH TO MY BRAIN."

I was born in the city of Medellin in Colombia in 1956. I had a very privileged background. My family was a very large and prosperous family that was established in this part of Colombia since the late 1500s. It was a very warm family because it was a very large and extended family—so many cousins and uncles and aunts. Medellin was a very different city. Back then it was about a half a million people, a provincial capital, very safe. This was before all the issues with drugs. In the 1950s and 60s, I started taking public transportation when I was in first grade. At age seven, I could go out, get on a public bus, and go to school. It was very safe.

I did all my primary and secondary education there and then started at the University of Antioquia to become a biologist, but it was the 70s, so there was a lot of political unrest. There were many episodes of the army coming into the campus and students getting shot and detained . . . semesters were cancelled. I was only completing about a semester per year. After several years of this, I got tired of it, so I transferred. I moved to attend a Jesuit university, the Xavierian University in Bogota. That's where I completed my undergraduate degree in biology, which is more or less the equivalent of a US Bachelor of Science in Biology.

Moving to the United States of America
I was ambitious in terms of learning more science and becoming a scientist, and I felt that the United States' universities had a wonderful tradition of scientific research, great facilities, and resources. I was excited about that possibility, and I was also not particularly excited about looking for jobs in Colombia for a number of reasons. I thought that, as a biologist, my research opportunities would be very limited. Most of the positions that existed were about applied work and

101

doing things like working for a tobacco company to increase the consumption of tobacco in Colombia, which I didn't want to do. There was a possibility of doing ecology, but I waited for about a year for an appointment from the government, and they said, "It's coming, it's coming, it's coming," but it never came. I said to myself that the odds of getting something done in that environment were very limited. I started to read more about the brain and neuroscience, and I became interested in neuroscience. I learned about a Colombian connection: a famous scientist, Rodolfo Llinas, a professor of biophysics at New York University and a very well-known neuroscientist. I contacted him, and he talked to me and offered me some guidance, and that's how I decided to apply to NYU for a master's program.

When you leave, you leave half your heart and more than half your soul behind. You lose relatives, family, the comfort of culture, and the sense that I know this place so well. I know how to navigate it . . . I'm completely comfortable . . . and I like all the aspects of the culture—from the countryside in the Andes to the local food to the sports scene—but it's different. Friendships are different . . . this is a different culture . . . and you leave all of that behind. You leave your parents . . . you leave your siblings and other extended relatives . . . You leave your language.

I made a conscious effort when I moved to New York to limit my associations with other Colombians because every time I met with them, I realized they are all talking in Spanish, and I wanted to improve my English. But you do lose Spanish, which is a beautiful language. Eventually you dream in English, and I love English. I love its powers and its potential to do certain things that do not work sometimes as well in Spanish. But I also feel the other way around about Spanish. There are certain things that I can express much more accurately in my native language. I process my native language, my mother tongue, in a completely different way. It goes through a different path to my brain . . . it feels completely different . . . and I feel, almost, that when I speak in Spanish, I become a different person. You leave all that behind. So, it's a huge loss. And you get a new world. You gain a new home, new friends and family. For me that's also powerful. And you gain a professional life, which I wanted.

I remember some of the details. I left either from Medellin or from Bogota. I landed at Kennedy Airport late at night. It was snowing. The city was a bit chaotic. I really couldn't get to the city but by cab. I remember Billy Joel was playing on the radio, and I made it to the Washington Hotel on 23rd and Lexington. It was not the best part of town then, but it was a very cheap, cheap, cheap hotel that no longer exists. You can Google it, and you will find the page for it somewhere because it's a place that had—in its very distant past—some distinguished customers. Within a few days, I could not afford the hotel. I realized I wanted to stay a few extra days, and I couldn't afford it, so I stayed at the 34th Street YMCA, which was much cheaper and much dirtier, but it was affordable. On the first day, I remember I stepped out of the hotel onto Lexington. I looked up and down the street. The first voice I heard was someone talking in Spanish. I looked at the city, and I said, "I love New York." I always have thought that it is an incredible, exciting city that offers so much—that was my first impression. I couldn't afford to live in Manhattan. When I was a student at NYU, I lived in the Bronx. I had to walk through neighborhoods that looked like war zones where buildings had been burned . . . it was more than a bit dangerous. I did that for about a year and a half before I applied and was accepted at Columbia University.

STEM Education and Career

Starting at NYU was very challenging, partly because my study habits were not in good agreement with what was expected of me as a student. Part of this was due to the fact that my English was not particularly good. I struggled to understand things . . . I struggled to perform to expectations . . . I was not able to really represent my own knowledge. They asked me if I had ever done research. I thought research was a very fancy thing that all the other students had done. Only much later did I realize that the things that I did at my college in Bogota were comparable to the things that undergraduates at good universities do in the United States.

So, it was very challenging. In some ways, my first experience as a student was a failure because I was applying to this program run by this well-known Colombian scientist, Rodolfo Llinas, and basically he told me, "If you don't make an A in every course, we will not accept you." I didn't have the study habits or really the ability at that moment to make an A in every course, so it seemed to me they're not going to accept me in this program.

After about a year at NYU, I got a teaching assistantship, meaning I was a laboratory TA. They started to pay me, and I was able to make ends meet. Before that, it was not clear whether that would happen and you live with the day-to-day uncertainty—uncertainty about what's going to happen tomorrow or in a few weeks, which shapes how you commit to work and how you are able to study. It was very hard. Sometimes you romanticize it—you think it was fun and so on and so forth, but it was also stressful.

I was not always interested in STEM. My main passion and interest as a child was history. I was precocious: I taught myself to read and write in kindergarten. At age seven, my father took me to the Continental Bookstore in Medellin. He talked to the owner of the bookstore, Mr. Vega, and told him that he wanted to open an account for me. From then on, I could go to this wonderful bookstore, and any book that struck my fancy I could buy. The fact that it that was available to me was a privilege. It was wonderful. I was not a good student in high school. I just didn't study because I was reading too much history and fiction. I would skip class, go into the school's wooded area and read, so I got into trouble. But when I was fifteen, I started taking chemistry and physics, and all of a sudden my grades improved. I saw a lot of my friends struggling with chemistry, but I thought it was not only easy but also fun and interesting. I then found it more interesting than history or than other things that had occupied me, so I started to pay more attention to the sciences. I thought about studying chemistry, knew it was not physics, but it was biology that ended up winning.

So how did I get here? It started with school at NYU, where I struggled, but my English improved after about a year-and-a-half. I took demanding classes, and worked in a research lab under Dr. Bernardo Rudy. When I entered Columbia, I was not at a disadvantage anymore, but in fact, at an advantage because I had taken one year and a half of graduate classes, which helped my PhD program go very, very well. I published in top journals. I also met my wonderful wife, Pam. That, to me, is the best thing that has happened to me in the US—that, and our children.

When we finished our PhDs, we decided to go to California to do our postdoctoral fellowships. I was at the University of California, San Francisco and Pam was at Berkeley. We were there for a year, and then my lab decided to move to the University of Texas in Dallas, so we lived in Dallas for four years. In Dallas we had our first child, Sara. At the end of our postdocs I started to look for a tenure-track position as an assistant professor. I applied for several positions, and I had several offers (and a good number of rejections). It was a hard decision, but I accepted a position as Assistant Professor of Physiology at the Emory Medical School in Atlanta.

I was at Emory for six years, when my tenure clock arrived. I needed to decide on whether or not to do tenure. My lab was well-funded; my work was well regarded; and I thought I was reasonably happy with the work. But I was not particularly happy with the life, the lifestyle, of the full-time research scientist. Research absorbed all my time, and it impinged on my family life (our son Eduardo was born in Atlanta). I was really busy, and Pam was also a very busy research assistant professor. Though there was not a lot of teaching in the position, I was teaching some. I found that medical students are wonderful people, but they were totally overwhelmed by the amount they needed to learn. Because of this, they usually could not really explore things in creative and playful ways. I decided that I wanted to do more teaching and decided to change my career a little bit.

I applied for jobs at several liberal arts colleges. Carleton College invited me for an interview. People thought I was crazy to leave a tenure-track job. My department offered to put me up for tenure, and my research grant had been renewed, so the prospects were good. But I walked away from it all and started from scratch at Carleton. It was a wonderful decision because I have been very happy. It's a fantastic institution. I have great colleagues and wonderful students. I teach, and I do some research. I have students working in my lab, and I mentor them, but my focus and emphasis lean more towards teaching than research, which is the inverse of what I did at Emory, but I think it's something that suits me better.

I don't really know whether there are advantages being an immigrant working in the field. I mean, there are a lot of non-native postdocs. I think there can be less interest by US natives to go into science, but a lot of interest in other parts of the world. I think that's part of it. At every academic institution I went to, there were lots of postdocs from Asia, Europe, and Latin America. In most places, I enjoyed an amazingly egalitarian culture. Columbia to me was a wonderful example, and I think NYU was the same given the multinational faculty. When all the differences melt away, it's just a learning community of colleagues with common interests. It's not hierarchical—if a first-year student has a great idea, it is a great idea. The great idea doesn't have to come from the famous, established professor. There's quick recognition for merit and originality. And in that sense, I never felt that I have been discriminated against. There have been other times in the US when I've felt uneasy about my role and my position . . . how I felt about myself as an immigrant in terms of how I relate with other people and so on . . . but usually not at the universities. I have always found them to be highly egalitarian.

Contributions and Impact

Carleton College has a wonderful neuroscience department. I have made a number of good friends in the neuroscience community in Minnesota, which has helped me be a better neuroscientist. I think the main thing that I learned were new ways to become an effective STEM teacher, instructor, mentor, professor, and guide to students working in my laboratory. I would say that my contribution in Minnesota has been more towards helping shape a future generation of researchers and professionals, mostly physicians. My contribution, I think, is not so much in terms of making new discoveries but as a professor. I've had my lab, and I've had some interesting findings, but these are minor as compared to helping shape the future generation of researchers and physicians.

Many of my students have gone to medical school. They send me postcards and emails from medical school and sometimes from residency saying things like, "I'm studying this area now . . . I'm working on this area now . . . I'm ahead of the curve . . . I'm so glad that I had you as a professor at Carleton." Also, many of my students have gone on to PhD programs in neuroscience and in other areas of biology. From them I hear comments such as, "I learned how to think about certain things from you . . . how to be an effective speaker . . . how to be careful in the laboratory." These are wonderful and rewarding things for a teacher to hear.

I live in Northfield, Minnesota, south of the Twin Cities with a population of 20,000. It is a small historic town on the Cannon River, which was used for grain mills. Northfield produces breakfast cereals at a large cereal factory. It has a strong rural and agricultural history, but it's also a college town with Carleton College and St. Olaf College. Carleton is a wonderful liberal arts school of about two thousand students with a wonderful faculty and lots of resources, including a significant financial endowment. It recruits nationally, with students from all states and around the world. The campus is beautiful: one of the most wonderful things is the Carleton Arboretum, which is about eight hundred acres of forest and prairie by the Cannon River.

I live four blocks from campus, so I often bike to campus during the better weather, and it's less than ten minutes on foot. If I have a class, I can walk out of my house ten minutes before class, and I am at the classroom on time. It's a small community, but the fact that it is a college town changes things because there's lots of people with PhDs. And because liberal arts colleges cover broad curricula, there are people from all fields—the sciences and also the arts, humanities, and social sciences. It's a very well-educated town, which creates a special community. It's a very enjoyable place, very low key, very safe. It's idyllic—a little bit like Lake Wobegon with two colleges. We really like it.

I never lived in a small town—Medellin had a population of two and a half million, then there was Bogota, New York City, San Francisco, Dallas, and Atlanta. I had never lived in a town as small as Northfield, where everyone knows everyone. That changes how you relate to people. You realize that you're going to meet people from all walks of life, and they're going to be your neighbors, and you have to get along. You cannot hide. My wife, Pam, has lived mostly in big towns. She loves Northfield and doesn't want to leave.

I would say without a doubt, that making people better physicians and better scientists through my teaching is my main contribution to the STEM fields, and the one that I'm most proud of. I think it overshadows any original scientific contribution that I might have had. Things that you think are very important, some may not see them as important, and things that you might even overlook, others might find interesting. I occasionally still get emails about old articles that I thought people were not interested in anymore, but someone is still interested. It is very hard to judge your own contributions, but it's not hard to judge your contributions as a teacher because your students remind you of it constantly, not necessarily by telling you but because you see their successes. And you see in their successes some of the things that you worked hard to instill in them and that you think you succeeded in.

greencardvoices.org/speakers/fernan-jaramillo

Ngarchelong,
Palau

OCEANIA

Simeon Ngiratregd

From: Ngarchelong, Palau **Field:** Information Technology
Current City: Linwood Township, MN

> "IN THE VILLAGE WHERE I GREW UP, WE INVENTED THINGS BY HAND. WE BUILT OUR TOOLS FROM SCRAP METAL AND FROM SPARE TIRES AND TURNED THEM INTO OUR SPEARGUNS. YOU BECOME CREATIVE, INNOVATIVE FROM MAKING SOMETHING OUT OF NATURAL MATERIALS; THAT'S WHERE I GOT MY DESIRE FOR INNOVATION."

The name of my home is called Ngcherau. It's quite important to know the name of your home because there is no address for the homes. It's called by a name and that home stays with you forever. My small village of about fifteen homes, Yebukel Ngarchlong, is very peaceful. There's no electricity, no running water. Growing up, my father was a fisherman, and my mom was a gardener, so as a child I learned all the tricks of the trade for fishing and gardening. I lived with my parents in this small village from first grade to sixth grade. Then I moved to the bigger island, Koror, which is the capital of Palau. Here, I stayed with my sister while attending high school.

I would stay in Koror throughout the school year, and then in the summertime, I would take a boat about thirty miles back to the village. Because of the slow speed of a diesel engine in the putt-putt boat, the ocean tides, the current, and the wind—the ride could take anywhere from six to eight hours. I would stay with my parents, and at the end of the summer, I would take the boat back to live with my sister in the big cities. I went back and forth until I was nineteen, then I decided to move out of Palau. One of my high school teachers, an American music teacher who grew up in Minnesota and attended Macalester College, helped me apply to Macalester College. My parents could not afford to send me to go beyond high school, I thought high school was the end of my schooling and then I would just stay in Palau and find a job, but I was fortunate to get a full scholarship. I believe my teacher and his wife were Peace Corps volunteers and became contract teachers when their terms were completed. Peace Corps volunteers came to Palau around 1966 or 1967.

Moving to the United States of America
After I got accepted to Macalester, I started preparing for my move. I needed

an orientation of what it's like to leave the country for the first time and what Minnesota is like. There was a tuna ice box, which is where the fishing industry stores frozen tuna. One of the men there would open this icebox and tell me that that's what Minnesota looks like—I didn't believe it. Now I realize that it was really true. I was very fortunate. My family was happy. As was the tradition, my father invited relatives from the village and had a little get together for my going away celebration. There was lots of food from my family and money donations from relatives. I had 400 dollars with me when I left Palau for Minnesota.

I started in the village of Yebukel Ngarchelong and took the little diesel engine boat about eight hours to Koror. From there, I took the airplane to Guam, but there was a stop at Yap Island. In Guam, I stayed with one of my brothers—I had three brothers and one sister that had already moved to Guam. Then I took a plane from Guam to Hawaii. I also had a brother who had lived in Hawaii at that time, so I stayed a couple days with him. He knew it was my first time away from home, so he just wanted to make me a little bit more comfortable. From Hawaii, I had a flight to Minneapolis with a stop in San Francisco. I remember that I almost missed that flight. There was a layover, and my English was not very good, so I was worried. It's kind of a bigger airport, so I couldn't find my way around. I just barely made my flight to Minneapolis. When I arrived, my host parents from Macalester met me with a sign welcoming me. They live in Bloomington, and they had three children. I was exhausted, so tired. They tried to show me the cities, but I was just wiped out.

One of the things I found difficult was the food, especially breakfast. On the island, we eat fish three times a day. It was hard to eat toast, eggs, juice, and milk. It's a totally different diet. Looking at obstacles, everyone spoke so fast. I learned English in high school in Palau, and I got a perfect grade in English grammar. I knew how to write it, and I thought I knew how to speak it, but when I got here, people used slang. They would say, "Oh, that's cool." It just blew me away—I couldn't understand what they were saying. That was difficult for me. Giving up the culture, the food, the warm tropical weather of the island, and suddenly you don't know anybody around you. There are so many people and so many friends—and the country is full of wonderful people, but I felt so lonely at times. You can't really share your culture. The weather was also difficult. I came here in September, and then the weather started changing slowly, so I had a chance to practice wearing heavy winter jackets gloves and winter boots. It was beautiful seeing fall colors and the anticipation for snow. I remember going out of my dorm room catching big chunks of snow, feeling it with my bare hands, and putting it in my mouth trying to figure out how I would describe it. The only thing that comes close to it in Palau is the shaved iced sold as flavored ice drinks. Imagine the cold weather has turned rain into shaved iced falling down from the sky—that's how I described it to my folks back home. I had a hard time adjusting to the heavy winter jackets, gloves, boots, and the smell of radiating and forced air heat. I froze my fingers on a regular basis because I forgot to wear gloves.

Giving up my family was scary. I didn't know what was ahead of me. Saying goodbye, especially to my mother was tough. She didn't give me a hug because it's more of a tradition to hide your emotions. There was no Internet, no telephone—it was 1969. I wrote letters, but I didn't know if they were going to get there. The mail would go on a boat, and it would take thirty days. At that time, I would say that it's almost like you're saying goodbye forever . . . you just never know . . . just have faith in God.

In 1969, when I came to Macalester, the Vietnam War protests were happening. The anti-war protests were all over the campuses, and they were putting down the US military and elected officials, calling them pigs and treating them disrespectfully. It was a confusing time for me because Palau was saved after the second World War from the Japanese empire, which had occupied the island for thirty years and would have kept the islands in the dark ages. I grew up with a lot of respect for the US military, especially the Navy personnel. They brought country and western music and movies, and they were a boost to the local economy. In fact, in my senior year of high school, I used my sister's car as a taxi. US soldiers were the best customers—a one-dollar fare from them was equivalent to a couple of days with local clients.

STEM Education and Career

I studied math and physics at Macalester, and it was in the physics labs where I encountered computers for the first time—a big IBM mainframe that utilized a punch-card reader for data entry. I spent many hours in the computer room curious about how computers work. At Macalester, the international student program helped me meet people from all walks of life, from many countries around the world—Tibet, India, Japan, Germany, Hong Kong, Taiwan, Ethiopia, France, Iran, Argentina, Brazil, Chile, Iraq, Morocco, Malaysia, Cyprus, Turkey, Cameroon, Ghana, and Democratic Republic of the Congo. Experiences living with these students ultimately shaped who I am and how I perceive the world.

In 1972 or 1973, I decided to leave Macalester and petitioned to switch my green card status from a student visa to a permanent resident. To qualify for that, you need to have a sponsor petition for you. My brother, who had retired from the US Navy, lived in Guam. Thanks to him it took only a couple of months for me to legally stay, work, and go to school. Ultimately, I was able to work at night and pursue my electronics studies at Brown Institute in Minneapolis during the day. I worked as a busboy at the Saint Paul Hotel, at the 3M factory making trim molding part for Cadillacs, at Saint Paul Linen services, at a meat processing plant in west Saint Paul, and at the Lindberg Heat Treating Company in Hopkins.

Deciding to leave Macalester without a four-year degree haunted me for years. I felt like I had failed so many expectations from family, but it was my instinct that led me to leave Macalester for technical studies, and I'm so happy that I did. My time at Macalester was not a failure—most importantly, it helped me choose the right career path. I was awarded an associate degree in electronics technology with certificates in computer technology, communications technology, audio technology, and color television technology. I landed a full-time job at Lindberg working from 3:00 p.m. to midnight to pay for my apartment, food, and tuition at Brown. Then I went to school from 9:00 a.m. until 2:00 p.m. I started out at Lindberg as a temp with a single task of feeding a huge baking oven with metal parts to be baked for achieving specific metal hardness. I was one of the top employees and was offered the opportunity for promotion. I spent two years at the factory, and they wanted me to stay, but I finished my electronics degree and wanted to go.

The electronic degree program was set up for guaranteed job placement in the field of your technical preference. Computer technology was my first choice, so I accepted the job offer and went to work for Delta Systems Inc., a small, family-owned business that designed and manufactured digital electronic cash registers. I worked with hardware and software engineers building circuit boards, system assembly, testing, and field installations. There were many challenges with integrated circuits since they were new, and I had not had much experience with them. I remember the electrostatic interferences as the root cause of ninety percent of system downtime. I spent many hours in the lab finding ways to help solve static interference. I traveled to North Carolina and to Philadelphia to fix freezing issues related to interference from static electricity.

My research and development tasks included traveling outside of Minnesota for new system installation, user training, documenting the install process, and any issues to help improve the product. A ski resort in Jackson Hole, Wyoming, was one of our first big installations. I flew into Idaho and rode in a four-wheel drive pickup truck in the subzero weather to Jackson Hole. Here I was, from a tropical island, the only dark-skinned person with all these people wearing cowboy hats and boots who'd never heard of Palau island. They were very excited. They took me up the mountains, on the ski lift, all the way to the top to the highest elevation I've ever been.

Returning to Micronesia

After four years at Delta Systems Inc, my girlfriend got a job as a speech therapist in Guam, so I said, "Let's go." This was 1979. At that time, computer technicians were rare. I could go anywhere and find a job. I got a job with Pacific Data Systems. They supported all the military bases, Air Force and the Naval base. They were getting tons of money and computer service contracts from the government of Guam, local businesses, and the US military. It provided hardware support, including installation, and upkeep and repair for electronic cash registers and mainframe computers.

Wang word processors and mini data processing computers were arriving on the island, and we became the authorized sales and support company for them. I specialized in supporting the new products and attended a two-week technical training for Optical Code Recognition (OCR) devices in Tokyo and a one-month training in Honolulu for Wang computers. While in Honolulu, I met a Wang's technician from Chicago who provided information to join Wang if I decided to return to the states. Providing hardware support in Guam and the Northern Marianas provided me with an appreciation for how electricity plays a very important part of computer stability and its life cycle. The majority of computer downtime on the islands were power related—power fluctuations and outages.

I was very happy to be back in Guam. It's just 800 miles from Palau, so my parents were able to come. We lived there for two years. The first time my parents met my wife was 1972, we were not married. We stayed with them in their house in Palau where all eight children were born and raised—no electricity, and real tropical fruits (star apples, soursops, oranges, bananas, mangoes, and papayas) growing everywhere. My dad told me, "Your life is finished," meaning I'd brought somebody to their home introduced her to my family, aunts, and uncles that I was obligated to marry her.

We lived on the island for a couple of years, then my girlfriend went back to the states to attend graduate school at the University of Wisconsin, Madison. I felt obligated to stay on the island and help my family, which is very much part of Palauan customs and traditions. Again, I had no idea of whether I was coming back. I enjoyed my stay in Guam until my oldest sister told me, "I just want you to know that if you think you are obligated to stay here and make your parents and us happy, then you are wrong. You need to go where you think is good for you. You've already made us very happy and very proud of you." Long story short, she basically wanted me to marry my American girlfriend and not waste my time enjoying the island lifestyle that often leads to alcohol abuse and addiction. This was the most helpful advice I have ever gotten. My father was a heavy drinker and gave it up when he became a minister. Many of my brothers and uncles had illnesses that were alcohol-related. My sister was thinking I needed to get out of here. It was the best decision. I always thought that I needed to come home and help the family, but that was the turning point of where I was going to be. I often say that my original plan was going to Minnesota for four years, return to the island, get a job, get married and settle down. It did not turn out that way, I met her, and the rest is history. I left and came back to the States, got married, and we settled down in St. Paul. We have two children, a son and daughter, and a granddaughter.

Moving to the United States of America a Second Time

When I returned to the states in 1981, I accepted a position with Wang Labs as a field engineer assigned for supporting the First National Bank in downtown Chicago. Wang is headquartered in Lowell, Massachusetts, but they had offices throughout the US. I was part of a downtown branch office team which consists of six technicians dedicated to supporting the entire First National Bank. It was nice not having to travel outside the building, just riding the elevator up and down. I was amazed working in a big, crowded city. The bank had thirty thousand employees compared to less than twenty thousand total population of Palau.

Wang was exploding, so after one year I transferred to Madison where UW-Madison was one of our biggest customers. At the time, wireless technology had not been developed, so we connected all the campus offices through cable. They were using copper coaxial cable. It was very expensive and eventually got replaced by telephone wires. Wang utilized a centralized concept where there is a central processing unit (CPU) handling all the processing tasks and data storage shared by multiple workstations at the users' desks. Workstations were considered dumb terminals only capable of displaying data with an input keyboard. My responsibility as a field engineer involved setup, configuration, and installation of the entire systems, including workstation terminals and other peripherals such as printers, external disk storage, and magnetic tape drives, as well as remote communications capabilities.

We were in Madison for four years, then I got transferred to Wang Labs in the Twin Cities. Downtown Minneapolis was my assigned territory. I worked in the high-rise buildings that were connected by skyways—I just went from one building to the next. Dorsey & Whitney Law Firm was one of my biggest customers. All the law firms in IDS Towers and the rest of downtown Minneapolis office buildings businesses were flooded with Wang computers. I was now a senior system engineer at Wang Labs. The salesmen hooked me up with the people who bought the computers, and I partnered with the end users developing plans for successful installations. Tasks included a computer room built and inspected to make sure it met the system specifics, system setup, configuration, CPU installations, workstations, and all the peripherals. Disk storage of six-hundred-megabyte capacity was considered top of the line and with enormous physical size, like a washing machine. I would go out and set up and install the entire system, plus all the devices. I worked at Wang Labs for seventeen years.

I accepted a desktop tech support job with HealthPartners. They were in the beginning process of replacing their legacy computer systems based on similar concept as Wang with CPU and dumb terminals. My immediate challenge was to erase my previous knowledge and replace it with a PC running Microsoft Windows operating systems. During the early years of implementing PCs and Windows networking, we manually configured individual workstations, which became semi-automated when cloning software tools were developed which allow the ability to capture an image on one PC and deploy it to multiple PCs. My last project at HealthPartners was evaluating thin devices to replace the PC workstations. Thin devices have a small footprint, are less expensive, are easy to maintain, and are almost one hundred percent immune from computer viruses. The device was approved, and I was pleased to see the evolution of thin devices coming back to replace the expensive PC workstation.

I was at HealthPartners for just short of twenty years. I retired on Mother's Day 2018, and we moved to this wooded area, and I'm finding myself gravitating to the lifestyle that I come from—almost nothing, just woods. I drive a Bobcat doing landscaping and moving earth. I'm on my second chainsaw taking down dead trees and chopping firewood. Of course, my fishing boat is walking distance to the lake. It's like Palau, all the lakes. Working in the yard, I look at how the squirrels are behaving and how the wind is blowing and the sun providing clues for the best time to go catch walleyes, sunfish, and crappies. Back on Palau, fishing was a chore; it wasn't for fun. Here, bass fishing—catch and release, is a sport that I'm not interested in. I want to eat the fish, so I fish for good eating fish like sunfish, perch, crappies, and walleyes.

So how did a guy like me become so intrigued with this? Going back to the village where I grew up, we invented things by hand. We built our tools from scrap metal and from spare tires and turned them into our spear guns. You become creative, innovative from making something out of natural materials; that's where I got my desire for innovation. My dad used to drag me through jungles to harvest resources (such as bamboo and vines) for building a bamboo raft, one of the most important tools for fishing. Vines are for building traps for fish as well as crabs.

Contributions and Impact

In terms of contributions I've made, it's the international culture that I've brought to my small community in St. Paul and the friends that I am still in touch with. I think it promotes an understanding of how far you have come, and what brought you here. At the end of the day, it's the love that you share with your fellow human beings. The international community that we have embraced is just tremendous. My wife is the granddaughter of Swedish immigrants, my adopted son was born in Guam, and my adopted daughter was born in India. For ten years, we hosted high school students from all over the world and continue to integrate them into our family. My host son from Colombia with his beautiful American wife and daughter are moving to attend a graduate school in California. Another host son from Brazil has moved for a job opportunity. We've visited yet another host son in Düsseldorf, keeping in touch with our other host children from Palestine, China, Japan, Brazil, and Reunion Island. They all studied at Nacel Open Door in St. Paul.

I volunteered in my community to teach young people some of the sports, and I was a coach for my son's baseball. The little kids that I work with, they're grown up now, and I'm proud of them. I remember in our community the focus of our sports was it's okay to be bad or not to play well or to lose the game. I always taught my boys that it's not okay; you have to work hard. If you work hard at it, if you practice, if you do your best, happiness comes from that. Because if you don't, that's where you're not happy. I hate to use the word failure, but successes come from hard work. It's something that I brought with me from growing up in this small village where you had to do things on your own. Where there's no electricity, no running water. You have to learn to take care of those things from a very young age. It comes from hard work. Commit and dedicate yourself to do a good job. I think I bring that to the small community that I live in.

I became a citizen in 2013. I went to my ceremony, raised my right hand. "Do you pledge allegiance to the flag?" I give up my rights and allegiance to my previous country to become a US citizen. That was tough. It really hit me—

brought tears to my eyes. Immediately after that, the judge who led the ceremony said, "Let me be the first to welcome you to the United States of America, and don't worry about the part that says giving up the country you came from, that's only politics. We encourage you to bring your values that make you who you are." We all come from a different background, but when we treat each other with respect, our cultures contribute to the goodness of mankind.

greencardvoices.org/speakers/simeon-ngiratregd

EUROPE

• Patras, Greece

Apostolos P. Georgopoulos

From: Patras, Greece

Field: Neuroscience

Current City: Minneapolis, MN

> "WE GOT THIS INSTRUMENT THAT TO ME WAS A DREAM . . . IT CAN LOOK AT THE HUMAN BRAIN. IF YOU KNOW HOW TO INTERPRET IT, REALLY YOU HAVE A STORY. IT WAS JUST FANTASTIC. IT WAS A DREAM COME TRUE."

I was born in 1944 in Patras, Greece, and I grew up there until I was fifteen. My father was a priest and left very early in the morning at 5:00 a.m. to go to the parish. Patras was a nice town. I grew up like a typical kid, playing lots of soccer; in fact, I still have a lot of scars on my knees from playing on unpaved roads.

A person lived with us who was my mother's uncle, who actually was an immigrant from the United States. He had come to the United States with his brother back in 1906. They worked in Washington state in the construction of railroads, then he served in the first war. He didn't marry. In 1939 he came to live with us in Athens when my mother was married. So I grew up with him living with my father, mother, older sister, and me. In fact, the first glossy magazine that I remember was The American Legion because he was a veteran and was in the American Legion. It's so funny because I currently hold the American Legion Brain Sciences Chair at the University of Minnesota Medical School. So, it is a very vivid memory—it was my first connection to this country.

At some point, when I was fourteen, my family moved to Athens from Patras. I was heartbroken because I had left all of my friends. I had a very socially active sort of environment, when I was a growing boy, and then we moved. It was painful, but in the long run it was very nice because I grew up in a bigger town, and then had a lot of ways to see what I wanted to do in the future.

STEM Education

I was very good at school—in other words, at the top of the class. It wasn't really anything special to me: I was just reading hard and doing well. But there, unlike here and other countries, in Greece you had to decide what to do next at the end of high school. It was six years of elementary school, six years of high school. There was no college, so by the end of high school you had to decide what to do. You can go into a technical career, you can become a scientist, or become something else. The highest regarded profession was engineering, and I didn't like engineering whatsoever, so that was out. I wanted to be a scientist to discover things, do research, and so on and so forth, but a career as a scientist in Greece was unthinkable at the time. The best you could do was to get an undergraduate degree specializing in physics (or mathematics or chemistry). The best you could do was to be a teacher, but I wanted to do more science. Another profession that was very prestigious that I didn't even consider was to become a lawyer. That was very money-making and had a lot of prestige, but I didn't like being in law school. So by exclusion, I went into medicine because it was the only chance to do a scientific career.

When I went to the national exams, which were very competitive, in my class there were 150 students, and out of three or four thousand applicants, I was admitted successfully. That was a very interesting transformation for me because it was a six-year medical school and didn't have any prerequisites for college, pre-meds, or stuff of that sort. During that period, at the end of the medical school, I was very interested in internal medicine, so I almost went to become an internist. What had happened in between was that by the end of the second year of medical school, if we did well in the exams, we could go and work as sub-assistants in unpaid student positions. I went to the Department of Physiology.

I liked the mechanisms very much. My heart was really interested in how the body works, the mechanisms—you think about how the heart pumps the blood and make it flow, how the kidneys filter, how the brain works, all of this stuff. So I went in, helping as a student but ended up as a MD/PhD graduate student. At the end of my medical school, I attended for another year so I could also have a PhD in physiology, and that was very helpful. In that way, I was exposed to science a lot. I studied biochemistry, and my thesis was in lipid metabolism.

It was in the third year of medical school that I met my wife in the physiology laboratory. She's an endocrinologist, a physician, working in both science and medicine. Through her I was exposed much more to the clinical aspects of medicine. Her interests complemented mine, and by the end of the medical school, it was almost like I flipped a coin and decided to do just research, which I am still doing. So that takes me to 1968 or so. That was when the medical school was finished, and in 1969 I got my PhD, and then I went into the army.

In Greece at that time, serving in the military was compulsory. I went to the army and spent two years there doing research in microbiology. When I finished, I had more publications in microbiology than in anything else. It was a very exciting time. As long as the problem or the question is interesting, I love it, and it doesn't matter whether it's economics or medicine by itself—research is just very exciting. These were very nice times because I served the two years in the Greek Army Medical Research Center, which was a research center that was actually making vaccines for meningitis. In addition, we were looking at other diseases that were more endemic in the Greek Army, like brucellosis. But what was most exciting at that time was that we had some outbreaks of cholera in the neighboring countries, and so we were charged to manufacture a vaccine for cholera, and I was very much involved in that effort, and it was so exciting to test the vaccine and assess its effectiveness.

I also had two very interesting experiences in terms of research. One was that the center I was serving at during my tenure moved from an old place where it was housed to a new hospital. During that move in the basement they discovered a lot of unopened cultures of salmonella that the person who established the institute in 1928 had collected. These had not been opened for forty years, and nobody knew whether the bacteria were living or were dead. Indeed, we found that they were alive, and we published a series of papers on that finding.

So, I didn't exactly go into microbiology, but at the end of these two years, I was thrilled with that kind of work. I had other interests as well. I wanted to become a psychiatrist to begin with. That was one reason I went into physiology, just to get my biology training before I embarked on psychoanalysis. That's what I wanted to do. But by the end of the medical school, I was not so thrilled with psychoanalysis and telling people what is right for them or not, and I had already acquired a much more biological background. But even research-wise, it wasn't clear what I wanted to do. I even applied to Guelph, an agricultural institute in Ontario, Canada, to do agricultural research. I also applied to a lab in Sweden to do brain biochemistry work and to another one in England to do endocrinology research. I even applied to the Carl Jung Institute in Switzerland to do Jungian psychoanalysis. It was just a very fertile sort of time with many possibilities open.

I even wanted to become a film director. I loved films! In Greece we had these open theatres, summer theaters, and you could go and watch old films. It was very inexpensive. The whole summer there was no school, and every other day, practically for nothing, you can go watch movies. And movies I loved, so I wanted to be a director!

Career in Neuroscience

I got married at the end of my medical school in 1968. Then we had two children: one was born in 1970, and the other one in 1972. After all of this thinking and up and down, I decided to do research in neuroscience. That was the best, as it mixed together my interests in psychiatry, psychology, and the brain. So, in 1969 I planned to explore fellowships in the United States, which was really the only very good option at the time. I wrote to three established brain scientists inquiring for the possibility of a research fellowship. I never got a response from the first; I was turned down by the second; but the third (and best) offered me a fellowship on the spot. He was Vernon B. Mountcastle at the Department of Physiology of the Johns Hopkins School of Medicine in Baltimore, Maryland. He said it was his dream to have a fellow from Greece; he told me in 1969, "I'll have a position in 1972 for you." That worked out because 1971 was the time when I was being discharged from the army. In fact, he and his wife visited Athens in May 1971, so we met, and I gave them a tour of the museum and Acropolis.

On October 22, 1972, we first set foot in this country. The kids were born in Greece, two daughters—the four of us were all Greek citizens. It was a nice arrangement because both my wife and I were assistant professors at the University of Athens in the Department of Physiology and took a two-year leave of absence to do our postdoctoral fellowship training.

We flew through London and boarded a Pan Am Boeing 707 bound to Baltimore. Here we landed in the fall of 1972 in Maryland. What a glorious scene! The leaves were bright red and orange and all colors in-between, and this was a huge expanse. That really was unbelievable. We thought all the houses would be like in New York, like you see in the films, one next to each other. But this was rural Maryland, and it was beautiful. It was just a revelation. We couldn't believe our eyes that this country was so beautiful. We stayed in Dr. Mountcastle's home outside Baltimore for a week. He found an apartment for us, a furnished apartment, where we stayed for a month. And then we found a place to live out in the countryside and started work.

It took some time to realize that American and British expressions can be quite different. We had learned English in the British style. So when you say "thank you," your typical response is "don't mind, don't mention." I had never heard the expression "you're welcome." We come here and everyone says "you're welcome." In the beginning I thought people knew we came from abroad and they were welcoming us. That was so funny. The environment, the place, Johns Hopkins, was just the best in the country if not the world, and we had a very fruitful two-year fellowship. But everything comes to an end, and in 1974 we left and went back to Greece.

It was November 30, 1974. I was so homesick; I was kissing the ground of my country. In February the next year we went for an excursion down to the sea, and I swam. I had just missed the sea so much. I almost became paralyzed because of the cold water, but it was just incredible. For better or worse, it didn't work to stay in Greece. Students were demonstrating all the time, and hardly anyone was attending our lectures. It was a very difficult period. What I wanted to do was research, so we decided to immigrate to this country in 1976, and we went back to Johns Hopkins and returned to our previous workplace. In 1978 I established my own laboratory looking at the brains of monkeys while they were making movements in different directions and recording the activity of brain cells in the motor cortex, which is the area that controls movement. Then two things happened essentially. One is that I made the discovery of how the brain processes the command to move your arm, i.e. what is the nature of the brain signals that will lead to the movements of the hands, and that paper we published in 1983. Now this became the basis of "neuroprosthetics," the field of using brain signals to control prosthetic devices. The other contribution was in cognitive neuroscience, which is how you interpret brain signals to decipher what is being thought of. So, my research in those years was a mixture of straight motor neuroscience and cognitive neuroscience.

In 1983 I became a naturalized US citizen. I had my interview with the immigration judge in Baltimore and passed the exams. Then he gave me advice. He said, "Young man, this is the time to think seriously about your future." I said, "What do you mean?" He said, "Think seriously about changing your name. You know with 'Georgopoulos,' you'll be lost. Clearly no one will pay attention to you. Why don't you call yourself 'Smith'? This is just the time to do it, when you get naturalized. Change your name to a true, common American name." I didn't, but it was interesting. It was offered as a sincere advice and in good humor.

I was a full professor at Johns Hopkins in 1986. My wife was an Associate Professor of Medicine, specializing in endocrinology and diabetes. That takes us to the end of the 1980s when I was offered to succeed my mentor, Dr. Vernon Mountcastle, but at the same time, we were looking at the other places. I had three offers: one at the University of Southern California in Los Angeles, another in New York University in the Washington Square campus, and the other here in Minneapolis. We decided on a new adventure and came here because, ultimately, everybody in our family preferred to come to Minneapolis.

Contributions and Impact

When we came here in 1995, we had one major decision to make. I was given a big offer to go to Harvard with all kinds of bells and whistles, but we decided to stay here, which was nice. Later it was a big discussion for me about going to Yale. For that we also decided to stay here. I think it paid off because we are very happy here, and the magnetoencephalography instrument that we have here is very high end—one of only a few in the world. So, we have been contributing to understanding not only how the healthy brain works but also how its function is altered in various diseases, including Post Traumatic Stress Disorder (PTSD), dementia, etc., and how we can differentiate one disease from another.

I have my wife and our two daughters that live here in Minneapolis. We have four grandchildren, eighteen years old to nine years old. We came from Greece with very little money, and we left everything behind—families, friends, homes. We came a family of four with no guarantee of financial support beyond a year, but we were fortunate to have been given the opportunity to prove ourselves and have succeeded in advancing science and medicine, which we still do.

greencardvoices.org/speakers/apostolos-p-georgopoulos

Esther Ledesma

From: Santo Domingo, Dominican Republic   **Field:** Industrial Design
Current City: Bloomington, MN

> "I'VE ALSO LEARNED TO NOT BE AFRAID TO BE WHO YOU ARE AND BRINGING THAT TO WHEREVER YOU GO. I BELIEVE ALL OF US HAVE SOMETHING TO BRING TO THE TABLE AND THAT BEING ABLE TO UTILIZE THAT INFLUENCE TO MAKE THINGS BETTER IS REMARKABLE AND HIGHLY IMPORTANT."

Some of my family, including my grandmother, were US citizens. When I was little, my parents obtained their green cards, and their intention had been to move to the US to pursue a better life. A couple of years went by and my parents separated. A few more years went by and I remember the moment our father communicated he was going to move. It was a Saturday morning. He came to our house wearing a blue blazer and said, "I'm moving, and I'm not coming back." While I didn't understand what that meant until years later, I now understand that in that moment my life became a "before" and "after" . . . and the after was very difficult. Although I was a child, I had to navigate what I needed to do not only for me but for my brother and sister. I wanted to make sure that they felt safe and that things would not be as bad for our mother and us as we thought they would be.

I think that people in the United States are disconnected from how difficult life is outside the country and they do not understand how difficult it is to immigrate here. When I was five years old, we went to the embassy to request a visa. It took us ten to fifteen years for the visa to come through. My mom said there would be a day we would leave, so I kept thinking what I had in the Dominican Republic wouldn't be there for long. In 2009, we finally got a date to come to the United States, but my brother, who was nine years younger than me, was not part of the initial request for the residency. My mother and sister made the decision to leave to the United States and reunite with my father, and I made the decision to stay in the Dominican Republic with my brother. I did not want him to feel the way I felt when our father left. I worked hard to get a scholarship to go to college, so in a way it worked out that I was able to stay with my brother and finish my college schooling there.

When I was in Dominican Republic, studying STEM careers was not necessarily an easy option for me, but I worked to earn a scholarship and get into one of the best schools to study industrial design. My mom was a teacher and always made sure we understood what education meant and how important it was for us to reach our goals. She wanted us to be whatever we wanted to be. I always have with me what she taught us about the importance of being educated. That understanding has helped me to grow and become who I am today. She's been my inspiration to keep going.

Moving to the United States of America

After I graduated from college with a bachelor's degree in Industrial Design, my brother had the opportunity to come to the US. I reunited him with our family, and I returned to the Dominican Republic. With my family in the US, it was hard not to have the immediate circle I grew up with or a house. For two years I basically lived out of my suitcase with extended family. I loved the work I was doing, but on December 22, 2012, I decided I wanted to be with my family, and it was time to make the jump. It felt like jumping into a pool and I didn't know if it had water in it or not. I quit my job and packed everything I owned into three bags. When I got to the airport, I couldn't take them all, so I had to leave things behind. I didn't realize how hard it would be to leave without looking back or knowing if I would ever return.

My family resettled in New Jersey. When I got there, it felt weird because I landed in a place I didn't belong. It did help to be around my family, but life was not easy. My family was living in a small apartment, and it wasn't a great situation. I did know how to speak English, and even though there are a lot of Dominicans and Latinos in New Jersey and New York City, the whole environment was not ideal. This was a shock because I felt I was put in an environment I was not immediately able to relate to.

STEM Career in the US

I had to quit my job to come to the US, so I was starting from zero. I remember once I got through the first week that I decided, "Okay, I need to figure out what I am going to do with my life." I started looking for jobs and began to interview. Sometimes I would have three interviews in one day. At one interview I was offered a job, but it wasn't in New Jersey. My parents felt like I had moved to the US to be with them, and now I was choosing to leave. I lived with them for only thirty days before moving to Massachusetts where I started working as an engineer. From there it has been this journey of non-stop evolving, assuming new challenges, and new opportunities. I've lived in five states since immigrating.

After working in Massachusetts, I was recruited by Eaton, my employer in the Dominican Republic, for a job in South Carolina. I accepted the position and found I was one of the few Latinas in town. Even if people were not very used to interacting with Latinos, their welcoming and southern charm made the situation easier. While I was there, I earned a master's degree in Product Development from the Rochester Institute of Technology. I was promoted and relocated to Pittsburgh, Pennsylvania. The next promotion brought me to Minnesota in 2018 as a global product manager at Eaton. Within these last two years, I assumed the role of a corporate co-leader for Eaton's Hispanic and Latino group at a national level, and was able to change the conversation around how we as a company are talking and embracing inclusion and diversity. More recently, I've assumed a Portfolio Marketing Manager role for Medtronic, where I'm playing an active role on how to deliver the best value to our customers and patients with our therapies and medical device products.

Contributions and Impact

During my years of work in STEM, I've learned that the representation of Latinos and Hispanics in the decision-making process matters. Whenever I get the opportunity to talk to young engineers, I tell them, "Hey, this is something you can do. There's somebody like you in a higher position that has gone through whatever you're going through." Seeing their faces and seeing how this can illuminate their paths is priceless. I've also learned to not be afraid to be who you are and bringing that to wherever you go. I believe all of us have something to bring to the table and that being able to utilize that influence to make things better is remarkable and highly important.

I am passionate about STEM and bringing a minority representation and standpoint to the table. I bring that to my work and to whatever organization I'm at. I enjoy working to bring people together towards a common goal. Additionally, I volunteer as much as I can. I am the President of the Society of Hispanic Professional Engineers (SHPE)—Twin Cities Chapter where I get to work and connect with local high school students, local professionals, and college students at the University of St. Thomas and the University of Minnesota. In addition to my professional and volunteer work, I am working on an MBA with an international business and marketing focus at the University of Wisconsin. And although the winter is not my season, I have embraced it as much as I can to make Minneapolis the place I call home.

greencardvoices.org/speakers/esther-ledesma

ASIA

Quetta, Pakistan

Zurya Anjum

From: Quetta, Pakistan

Field: Psychiatry

Current City: Sartell, MN

> "IN THE PAST, IT WAS A NOVELTY TO GO TO A PLACE WITH SNOW AND MAKE A SNOWMAN. YOU WOULD LIVE FOR A FEW DAYS THERE AND THEN GO BACK HOME. BUT DRIVING IN THE SNOW . . . BUNDLING UP TWO SMALL CHILDREN . . . THEN GOING OUT IN THE SNOW AND DRIVING? IT TAKES A LOT OF GUTS TO LIVE."

I was born in 1971 as the youngest of three daughters. My father served in the Pakistan Army. Even though education for women was not considered a priority in those times, my mother was very motivated to get a higher education. She had two master's degrees—one in philosophy and one psychology—and taught at the collegiate level throughout her career. Both my parents put a very strong emphasis on education for us. We moved a lot when I was younger because my father was in the army, and we would get stationed every two or three years in a different city. We would then pack up our whole house and move.

My father retired from the army when I was about eleven years of age, and we lived in the city of Rawalpindi for about fifteen years before I got married. Both my sisters were interested in becoming physicians, so they got into medical school and completed it. I was not interested in becoming a physician, but my parents encouraged me to consider it as a profession, and now I truly believe that this is the best profession for me. Getting admission to medical school in Pakistan is quite tough. After admission, it takes five years to complete the degree. You then have to do a year of internship and then pursue post graduate training if desired. I was more than halfway through my post-graduate training when I got married to my husband.

Moving to the United States of America

In Pakistan we have mostly arranged marriages, so I met my husband just a few days before the wedding. I had never seen him before or talked to him before, but my parents knew his parents, and they knew their family background, which was similar to ours in many ways. It was very hard to come to the US after getting married to someone I had just met. Moreover, I was leaving my country, my family, my profession, and all my friends. The one thing that was really helpful was the fact that I had an older sister who was also in the US living in the same state. Language was not a barrier for me as I was educated in Pakistan in a school run by Catholic nuns from Ireland, so my English was very good.

People often ask me what my first thoughts were when I came to the US. I think the first thing I remember when I came here was that my luggage got lost on the way and did not arrive. When I landed at the airport, I did not even have a toothbrush to my name! Everything that I had brought over did not come for at least three days. I felt so lost as they were the only things familiar to me and were a big source of comfort at that time. When I got here, my sister was the only person besides my husband that I would talk to daily. She would help me out a great deal with everything about life here in the US. My sister had been in the US for about seven years already and was working as a physician. It was hard to adjust to life in a different country, particularly since I could not drive or work here without first getting my licenses to drive and practice medicine.

At such times, small things have a big impact on your life. An excellent example of this is that I had been driving a car in Pakistan for ten years. I drove everywhere; I drove family members places where they needed to go. In Pakistan we drive on the left side of the road, and I was very comfortable with it. Switching to the right side of the road proved to be quite a challenge. I failed my driving test twice, and I was very disheartened at that time. I told my husband that I could not envision a life in this country. If I could not drive and had to depend on other people to drive me to places, I needed to go. Fortunately, the third time I passed. It made me think about how small things we take for granted once we learn them can turn into a big obstacle in our life when things change.

Similarly, I had completed my medical training back home and was a practicing physician. However, when I came to the US, my medical degree was not recognized by the US, so I had to take five exams to become qualified to work as a physician here. That was a very tough decision to make because it meant a lot of study, a lot of time commitment, and a lot of money. I felt very passionate about my profession and really wanted to pursue it here. I had the chance and the opportunity to study here and advance my professional knowledge. After taking the examinations over one year, I started my residency training for four years in psychiatry and completed it. I feel very proud of my accomplishment as it was not easy, but I persevered with the support of my family.

I had never lived in a place that had snow. In the past, it was a novelty to go to a place with snow and make a snowman. You would live for a few days there and then go back home. But driving in the snow . . . bundling up two small children . . . then going out in the snow and driving? It takes a lot of guts to live through that! During my first winter in Minnesota I would look out of our apartment window at a clear sunny day and think it would be a nice day outside. It took me some time to learn that it was actually the opposite. Clear winter days are extra cold!

Culture shock is another big issue. Everything is different—from the language to clothes to food and more. It takes a lot of effort before you learn to adjust to it. Financial hardship is another thing. Coming from a country whose currency is very low compared to the dollar, everything seems extremely expensive. Before buying anything—from food to clothes and any other thing— you are always doing that mental math in your head and trying to figure out if this thing you are buying is more expensive or less expensive? Is it worth it to buy, or is it not worth it? Eventually you get to the point where you stop doing that because you cannot survive doing that all the time.

Confronting Stereotypes

Before September 11, 2001, I used to go outside in my cultural clothes and people always commented on the design or colors in a good way. However, after September 11, 2001, people would stop and stare. They did not say anything but it made me uncomfortable, so I stopped wearing those clothes outside of home.

On the day of 9/11, I was on-call as a resident in a level-one trauma center hospital. Early in the day when it was thought that there would be a lot of casualties, we were told to discharge all non-urgent patients. In the evening, all on-call residents, including me (about twenty-five), were watching the news in the resident's lounge when we realized it was a terrorist attack by terrorists who identified themselves as Muslim. As the only Muslim in the room, I was suddenly extremely uncomfortable. What if someone said something to me? What would I say? Thankfully, no one said anything. It makes you realize how much you take for granted being a citizen of a country versus a first-generation immigrant to that country.

About a year before doing the interview with Green Card STEM Voices, I came to the realization that I needed to share more about my culture and religion. My fourth-grade son came home one day and asked if we were terrorists. I asked him where he'd heard that, and he said someone at school said all Muslims were terrorists, and since he was a Muslim, he was one too. That was my wake-up call to help people understand who we are as part of this community and nation.

People have so many stereotypes about religion. People ask me a lot of questions: why I wear Western clothes and do not dress like Somali people . . .

if I have to ask for permission from my husband to work . . . why I do not wear a hijab. Some go even so far to say I do not look like a Muslim! My response is always, "Do all Muslims have to look the same? Do all Christians, Jews, Hindus, or people from any other faiths look the same?" Those are stereotypes people believe in for all Muslims, so they expect everyone to follow them. A lot of these are cultural practices, and Muslims all over the world follow them differently based on their culture.

People are hesitant to ask questions. When I speak at churches about my faith, I get a lot of questions—sometimes strange ones. I got asked once why older Somali men have red beards. A lot of older Somali men dye their beards red. Why do they do that? It's cultural. Henna is a big thing in their culture. Women use of it on their hands and feet. Men use it on their heads and beards when they grey because it's natural and has been done in their culture for a long time. Henna is derived from a plant, so it is natural, and there aren't any chemicals.

Personally, I try to help out in my community, and I'm a very active member as a volunteer at both my kids' schools. Also, I help out when people ask me to speak at different occasions to help them to get to know our culture and our religion or to answer any questions that they have about how we live and what is different about us.

I think my life here now is pretty much the same life that most people live. I have a job and two kids that go to school. I'm running around most of the time to and from school or to their extracurricular activities. When my kids were in elementary school, I would volunteer at their schools frequently and got to know a lot of their teachers who really appreciated my help. Every year around the Muslim celebrations of Eid, I would talk to my kids' classmates in school and answer their questions about our faith and celebrations. Kids are so innocent and ask questions without any fear or prejudice. I believe that this is one of the most important things we can do to help each other recognize that we all pretty much have the same life. We're just living it a little bit differently.

STEM Career

Professionally, I have been working here now as a psychiatrist for almost fifteen years. I take great pride in this, and I feel very blessed to be able to work and help people deal with their mental illness. Although the stigma of mental illness here in the US is not as severe as it is back home, I am still surprised to see how many people feel embarrassed or ashamed about admitting that they have a mental problem. To be able to help them deal with that and get better is very fulfilling for me.

For the past three years, I've worked at the St. Cloud Veteran's Administration Hospital (VA). What surprised me when I started working there is the perception that when you go to the VA, patients are all going to be very old. Actually, the majority of patients are in their thirties, forties, and fifties. The older veterans are from the Vietnam War. The younger veterans have served in Afghanistan, Iraq, and other parts of the world as part of peacekeeping forces. The Reserves get called into all parts of the world. I was also surprised at the number of female veterans I saw—I thought it would be all men. It turned out to be very eye-opening and gratifying to not treat just a single segment of population but rather the whole age range. The St. Cloud VA is more specialized towards psychiatry. Minneapolis is the big hub for other specialties. For psychiatry, we have a treatment program for PTSD and chemical dependency with about 150 inpatient beds and also an acute psychiatric facility with fifteen beds. We also have an outpatient clinic program. People sometimes drive two to three hours one way to get here to be seen. Although it is inconvenient for them, they are very appreciative of the care they receive at the VA. Some psychiatrists at our VA use telepsychiatry to see patients who cannot drive out to the VA.

Contributions and Impact

Besides work, I am active in organizations that work to improve our community. I am on the board of FACT, which stands for Feeding Area Children Together. We help to raise money and provide food for about two hundred elementary kids in our area who do not have food at home over the weekends and during school breaks. We pack sustainable food like dry pasta, apple sauce, cereal, cheese sticks, and canned chicken and fish. Then we put the items in the children's backpacks before weekends and long breaks. The need has grown so exponentially that we're just trying to figure out what to do—the waiting list is so long. You hear about hunger in poor countries, but many people don't realize it's happening in our own backyard. To be living here and to have kids who are going without food . . . it just breaks my heart. I'm really proud of the work we're doing.

I've also been the co-chair of UniteCloud for the past year. This is an organization that is working to bring together people from different faiths, races, and gender identities. We do things like go to mosque every Friday during Ramadan to distribute free dates to our Muslim community. We celebrate Pride week, African American and Somali events, and other racial and cultural events. The organization's work is really good and is being recognized more and more, both locally and nationally. We provide diversity training and hold vigils when there are events locally and nationally that are racially or religiously biased. I also serve on the board of Great River Regional Library. My goal is to diversify the literature options we have for readers in our libraries and add a cultural perspective to the board.

St. Cloud has around fifteen to twenty families from Pakistan. It's a close and well-knit community. We celebrate events and holidays together and help out when anything is going on. I have been living in the US now for twenty years. When I came here, my husband was living in Minnesota. After going through the first winter here, I told him I am not going to live in this state. I had never lived in the snow, and although I had seen it on vacations, living daily through a Minnesota winter was a shock. Honestly though, Minnesota really grew on me. People are very nice and friendly. It is a very educated and work-friendly environment. There is something to be said about "Minnesota nice." Over twenty years' time, we have raised our two children here and have learned to love Minnesota, despite the weather.

greencardvoices.org/speakers/zurya-anjum

Afterword

After the publication of this book, our hope is to increase these stories' impact and amplify the storytellers' voices by creating engaging programming that continues to break stereotypes and increase knowledge of the immigrant experience. This programming includes: a traveling multimedia exhibit featuring portraits, quotes, bios, and links to the storytellers' video narratives; a speaker series with the immigrant STEM professionals where they share their expertise in-person; and a podcast series that focuses on their stories.

You can also further engage with your communities and learn more about contemporary immigration through *Story Stitch*, our card-based, guided storytelling activity that connects individuals across different backgrounds by encouraging them to share and connect through stories. In addition, the Green Card Voices' traveling state and national exhibits allow for a visual experience with our authors' journeys. These interactive exhibits feature QR links to video narratives and are designed to expand the impact of GCV's published collections of personal narratives. Finally, for all the podcast lovers—please check out *Green Card Voices, the podcast*. It's available on Apple Podcast, Google Podcast, RadioPublic, Spotify and other podcast platforms. Each bi-weekly episode features one first-person story of an immigrant or a refugee living in the United States.

Immigration plays a significant role in the US: one in five Americans speak a language other than English at home. From boardrooms to book clubs, from High Schools to Universities, from the corporate leaders interested in understanding their team to the individual interested in learning more about their immigrant neighbors, this book is an important resource for all Americans. We hope it will spark deep, meaningful conversations about identity, contribution, appreciation of difference, and our shared human experience.

Finally, we hope that learning from the authors' stories featured in this book will be just the beginning. The more important work starts when we engage in the difficult, essential, and brave conversations about the changing face of our nation. To learn more about speaking events, traveling exhibits, and other ways to engage with the Green Card Voices stories, visit our website: www.greencardvoices.org.

Glossary

Acropolis: a fortified settlement or citadel of ancient Greece that was usually built on a hill overlooking a city; the Acropolis of Athens is the most famous and contains the remains of several historically and architecturally significant monuments

ACT (American College Testing): a standardized test taken by high school students in order to be admitted into American colleges and universities

Center for Industrial Technological and Services (CBTIS 7): a program that allows high school students to obtain a certificate or technological baccalaureate degree while earning a high school diploma, located in Reynosa, Mexico

Citrix server: developed and distributed by Citrix Systems, Inc.; a tool for computers that allows businesses to host applications and resources all on one server so they can deliver those resources to clients quickly

Central Processing Unit (CPU): the electrical circuit within a computer where most of the calculations are processed and where instructions of a program are carried out

D train: a subway line in New York City, starting at 205th Street in Bronx, passing through Manhattan with a final stop in Coney Island, Brooklyn

DACA (Deferred Action for Childhood Arrivals): an immigration policy established in 2012 that allows immigrant youth who arrived in the US as undocumented minors to study and work in the US on a renewable two-year delay of deportation

DAISY Award: also know as The DAISY Award For Extraordinary Nurses is a recognition program that honors the super-human work nurses do for patients and families every day; over 4,000 health care facilities and schools of nursing in all 50 states and 25 other countries, are committed to honoring nurses with The DAISY Award

Diversity visa lottery: also known as the Diversity Immigrant Visa program or the Green Card Lottery; is a US government program that grants 50,000 US Permanent Resident Cards to immigrants annually, based on the results of a random drawing, and seeks to select countries with historically low rates of immigration to the US

Diwali: a spiritual festival of lights observed by Hindus, Jains, Sikhs, and Buddhists that takes place over the course of four or five days every year to commemorate the triumph of light over darkness

DMV (Department of Motor Vehicles): a branch of the government in each state that provides citizens with drivers' licenses and vehicle registrations

Dot-com bust: also called the dot-com crash; the crash of the rapid Internet-business growth that resulted in the closure and loss of stock for many online companies and business in the early 2000s

Düsseldorf: the capital of North Rhine-Westphalia, a German federal state located in western Germany with a population of approximately 600,000

DV visa: see "Diversity visa lottery" above

Eid: a religious holiday celebrated by Muslims worldwide that signifies the end of Ramadan

ELL/ESL (English Language Learner/English as a Second Language): English language study programs that educate students who speak a language other than English but reside in a country where English is the primary language

Endocrinologist: a doctor who specializes in studying, diagnosing, and treating diseases and disorders of the endocrine system, which is the system that control hormones

Estrogens: natural or synthetic hormones important to sexual and reproductive development; intensive modern agriculture and waste disposal systems can discharge pollution levels of estrogens into the environment and negatively impact human, animal, and plant-life

FBI (Federal Bureau of Investigation): a US government agency that investigates when someone breaks a federal law or when crimes threaten national security

FDA (Food and Drug Administration): a US government agency that oversees the manufacturing and distribution processes of food, prescription and over-the-counter medications, vaccines, medical devices, etc. to promote and protect public health

Gastroenterology: a branch of medicine focused on studying, diagnosing, and treating diseases found in the stomach, intestines, digestive tract, and organs

Gates Millennium Scholarship: an academic scholarship funded by the Bill & Melinda Gates Foundation that aids low-income, high-achieving ethnic minority students to earn a higher education

Geotechnical Engineering: a branch of science and technology focused on determining the safety and capability of building and designing structures to be constructed either on the earth's surface or underground

Green Card: a commonly used name for a Lawful Permanent Residence Card, an identification card attesting to the permanent resident status of an immigrant in the US

H-1B visa: a temporary visa that allows US companies to employ nonimmigrant professionals with a college degree to work in specialized fields such as IT, finance, engineering, medicine, etc.

Henna: a natural dye with rich cultural history in parts of Africa, the Middle East, and the Indian subcontinent; used to dye fabrics, hair, and beards reddish-brown and to stain skin with temporary designs

I-20: a temporary visa that allows students from foreign countries to participate in student exchange programs in the US and attend schools within the US

INS (Immigration and Naturalization Service): a former US government agency that oversaw the naturalization, asylum, and permanent residency of immigrants coming to the US; in 2003, INS was divided into three new agencies: US Citizenship and Immigration Services, US Immigration and Customs Enforcement, and US Customs and Border Protection

IOM (International Organization for Migration): an intergovernmental organization that provides services and assistance to migrants, including internally displaced persons, refugees, and migrant workers

Jungian Psychoanalysis: a form of therapy designed to allow the analyst and the patient to work together to help the patient feel balanced by merging the conscious and unconscious parts of the mind

Khmer: an ethnic group native to Southeast Asia, particularly Cambodia; today the majority of Khmer people live in Cambodia, Vietnam, and Laos; thousands of who resettled in Western Countries following the aftermath of the Khmer Rouge regime and civil wars

Lean Six Sigma: a collaborative methodology designed to improve workplace performance and remove waste in the manufacturing industry

Lipid metabolism: the process by which lipids, which are primarily fats and oils, are converted into energy for the body

Magnetoencephalography (MEG): a non-invasive neuroimaging technique that allows doctors to study human brain activity

MCAT (Medical College Admission Test): a standardized test used to assess students applying to medical schools in the US, Australia, Canada and the Caribbean Island.

MLA (Modern Language Association): describes the specific style and formatting often used in college-level writing

Nacel Open Door (NOD): a nonprofit organization that focuses on expanding international understanding and language education and offering academic programs in public and private schools

NAFTA (North American Free Trade Agreement): a treaty signed by the US, Canada, and Mexico in 1993, designed to prevent high tariffs and reduce other barriers to trade between the three countries

NCR (National Cash Register): an American technology company focused on machines that assist in transactions and banking worldwide

Neuroprosthetics: a branch of medicine and biomedical engineering focused on developing and creating devices to substitute for missing body parts by connecting the device to patients' brains in order for them to control the device

Neurotransmitters: the chemical in the brain and body that make it possible to move and regulate behavior and emotions

Open Source: refers to a type of computer program whose code can be found online and be legally modified to fit the user's preferences

OPT (Optional Practical Training): is a period during which undergraduate and graduate students with F-1 status are permitted by the United States Citizenship and Immigration Services (USCIS) to work for one year on a student visa towards getting practical training to complement their education; beginning on March 11, 2016, certain F-1 students who receive STEM degrees, and who meet other specified requirements can apply for a 24-month extension of their post-completion OPT, giving STEM graduates a total of 36 months of OPT

Optical Code Recognition (OCR): known more commonly as optical character recognition; a software that allows images of texts or PDFs to be converted into files that can be edited

Peace Corps: a service opportunity for motivated changemakers to immerse themselves in a community abroad, working side by side with local leaders to tackle the most pressing challenges of our generation

Peacekeeping Force: a force designated to to maintain peace and security, facilitate the political process, protect civilians, assist in the disarmament, demobilization and reintegration of former combatants; support the organization of elections, protect and promote human rights, and assist in restoring the rule of law

PSEO (Postsecondary Enrollment Options): US program that allows students in tenth, eleventh, or twelfth grade to earn college credit while in high school

R&R: an abbreviation that stands for "rest and recuperation" or "rest and relaxation";

most often used by members of armed forces to describe the free time and leave away from their usual duties

Ramadan: the ninth and holiest month in the Islamic calendar where healthy Muslims fast daily for a month-long from sunrise to sunset focused on prayer, reflection, and community; an occasion to become closer to Allah and strengthen the relationship with him

Reunion Island: a French overseas island located in the Indian Ocean and east of the island of Madagascar, with a population of approximately 900,000

SDRC (Structural Dynamics Research Corporation): a company specializing in automotive and aerospace tech design and assembly; non-operational since 2001

Student visa: a specialized form of visa required for internationals to enroll and study in American schools and is valid so long as the student is enrolled in school (see I-20)

Telepsychiatry: a branch of medicine that focuses on the field of psychiatry and allows patients to receive psychiatric assessment and care through telecommunication over long distances with technology such as videoconferencing

VA (United States Department of Veterans Affairs): a US government agency that provides military veterans with educational assistance, health care services, disability compensation, and other benefits

Visa restrictions: refers to travel limitations on visa holders based on their country of residency

Visa: an official government document that temporarily authorizes individuals to be in the country they are visiting

Wang legacy systems: processors designed and produced by Wang Laboratories to produce, manipulate, and read the text of a computer screen

War on Drugs: a US federal government campaign launched in the 1970s dedicated to eliminating illicit drug use in the US through escalated global military and police efforts against the drug trade; the campaign has been largely unsuccessful and linked to the rise of massincarceration in the US as well as the proliferation of drug-related violence around the world

WHO (World Health Organization): an international agency developed and run by the United Nations that is dedicated to public health

Work visa: legal permission to work in the specified country

About the Advisory Team

Tea Rozman Clark, PhD is the founding Executive Director of Green Card Voices. She is an NYU graduate in Near and Middle Eastern Studies and has a PhD in Cultural History, specializing in oral history from the University of Nova Gorica. She is a first-generation immigrant from Slovenia and 2015 Bush Leadership Fellow.

Julie Vang is a writer and the Program Manager at Green Card Voices. She graduated from the University of Minnesota, Twin Cities with a Bachelors in Family Social Science and a minor in Asian American Studies. She is Hmong-American with over six years of policy and programming experience.

Ellis Sherman is the Lead Graphic Designer at Green Card Voices. He graduated from the University of Minnesota, Morris with a Bachelors in Fine Arts. In addition to working at Green Card Voices, he is passionate about creating noetic visual narratives.

Lara Smith-Sitton, PhD is the Director of Community Engagement and Assistant Professor of English at Kennesaw State University where she works to implement service learning and community engagement projects for students. Her teaching and research areas include community engagement, professional writing, and 18th and 19th century rhetoric. She serves on the Green Card Voices Board of Directors.

Robby Callahan Schreiber is the Museum Access & Equity Director at the Science Museum of Minnesota. He works within the museum's STEM Equity & Education Division with the goal of intentionally engaging and co-creating community based programming with individuals and groups underrepresented or disconnected from museum experiences.

J. Roxanne Prichard, PhD is a Professor of Psychology and Neuroscience, and HHMI Inclusive Excellence Director at the University of St. Thomas. Her work includes the allocation of funding from a one-million-dollar grant to create a campus environment where inclusivity thrives through constant reflection, analysis, and accountability.

Richard Benton is an IT Manager at Ecolab and serves on Green Card Voices Board of Directors. He has led Ecolab's global mindset ERG, EcoMondo. Inside and outside of Ecolab, he has worked to hear the voices of immigrants for many years. As a passionate language-learner and advocate for learning the languages of our communities, he strives for a more open space where US culture embraces immigrants and the benefits that they bring.

About Green Card Voices

Founded in 2013, Green Card Voices (GCV) is a Minneapolis-based, nationally growing social enterprise that works to record and share personal narratives of America's immigrants to facilitate a better understanding between immigrants and their communities. Our dynamic, video-based platform, book collections, traveling exhibits, podcast, and *Story Stitch* circles are designed to empower individuals of various backgrounds to acquire authentic first-person perspectives about immigrants' lives, increasing the appreciation of the immigrant experience in the United States.

Green Card Voices was born from the idea that the broad narrative of current immigrants should be communicated in a way that is true to each immigrant's story. We seek to be a new lens for those in the immigration dialogue and build a bridge between immigrants and their community—from across the country. We do this by sharing first-person immigration stories of foreign-born Americans, and helping others see the "wave of immigrants" as individuals, with interesting stories of family, hard work, and cultural diversity.

To date, the Green Card Voices team has recorded the life stories of over 400 immigrants coming from more than 120 different countries. All immigrants that decide to share their story with GCV are asked a series of open-ended questions. In addition, they are asked to share personal photos of their life (in their country of birth and in the US) included in the video narratives, which accompanies each essay. These video stories are available on www.greencardvoices.org.

Immigrant Traveling Exhibits

All of the stories in *Green Card STEM Voices* are featured in a traveling exhibit that is available to schools, universities, libraries, workplaces, and other venues where communities gather. The exhibit has twenty story banners—each with a portrait, a 200-word biography, and a quote from each immigrant. A QR code is displayed next to each portrait and can be scanned with a mobile device to watch the digital stories. The following programming can be provided with the exhibit: panel discussions, presentations, and community-building events.

In addition to the STEM Voices exhibit, Green Card Voices has ten other traveling exhibits focused on different themes and local communities across the Midwest and South. Our National Exhibit travels across the country. To rent an exhibit, contact us at or 612.889.7635.

Green Card Voices
2611 1st Ave S.
Minneapolis, MN 55408

612.889.7635
info@greencardvoices.org
www.greencardvoices.org

Order Through Our Distributor

Our books and Story Stitch are distributed in the US & Canada by Consortium Book Sales & Distribution, an Ingram Company.

For orders and customer care in the U.S., contact:

Phone: 866.400.5351

PCI Secure Fax (orders only): 731.424.0988

Email (orders only): ips@ingramcontent.com

Online: ipage.ingramcontent.com

Electronic orders: IPS SAN 6318630

Mail: Ingram Publisher Services   Attn: Customer Care
1 Ingram Blvd., Box 512, La Vergne, TN 37086
Hours: Monday–Friday, 8:00 am to 5:00 pm CST

For orders and customer service in Canada, contact:

Phone: 800.663.5714
Email: orders@raincoastbooks.com

Now Available:

Story Stitch

Telling Stories, Opening Minds, Becoming Neighbors

Story Stitch is a guided storytelling card activity that connects and builds empathy between people of different cultural backgrounds. It was created by the diverse Minneapolis / St. Paul community in a series of co-creating game sessions lead by the Green Card Voices team. This card game is perfect for: classrooms (ages 10+), diversity trainings, workshops, work places, leadership/fellow retreats, conferences, elderly homes, and more. Available as a deck (ISBN: 978-1-949523-11-9).

Contents:

- 56 color laminated cards total
  - Story Cards: 33 cards with story questions
  - Stitch Cards: 23 cards with lines and dots
- 1 four-sided accordion of instructions

Green Card Voices Store

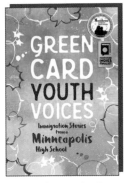

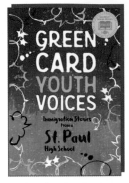

Immigration Stories from a Minneapolis High School
ISBN: 978-1-949523-00-3

Immigration Stories from a St. Paul High School
ISBN: 978-1-949523-04-1

Immigration Stories from a Fargo High School
ISBN: 978-1-949523-02-7

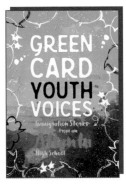

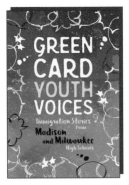

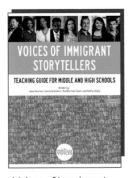

Immigration Stories from an Atlanta High School
ISBN: 978-1-949523-05-8

Immigration Stories from Madison & Milwaukee High Schools
ISBN: 978-1-949523-12-6

Voices of Immigrant Storytellers Teaching Guide for Middle & High Schools
ISBN: 978-0-692572-81-8

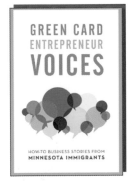

Green Card STEM Voices: Stories from Minnesota Immigrants Working in Science, Technology, Engineering, and Math
ISBN: 978-1-949523-14-0

Green Card Entrepreneur Voices: How-To Business Stories from Minnesota Immigrants
ISBN: 978-1-949523-07-2

Story Stitch: Telling Stories, Opening Minds, Becoming Neighbors
ISBN: 978-1-949523-11-9

Purchase at our online store: *www.greencardvoices.org/store*

*By buying this book you are directly supporting
the mission of Green Card Voices.*